Mathematical Analysis

Mathematical Analysis

Editor

Lavra Filipek

scitus
academics

Mathematical Analysis
Edited by **Lavra Filipek**

ISBN: 978-1-68117-258-3
Library of Congress Control Number: 2016934801

www.scitusacademics.com
Box No. 4766, 616 Corporate Way,
Suite 2, Valley Cottage,
NY 10989

Printed in United States of America on Acid Free Paper

PREFACE

The study of limiting processes is the central theme in mathematical analysis. It involves the quantification of the notion of limit and precise formulation of intuitive notions of infinite sums, functions, continuity and the calculus. In a rough division of mathematics, mathematical analysis deals with inequalities and limits. In some of its branches, such as asymptotic analysis, these aspects of the subject matter are readily apparent. In others, such as operator algebras, they are concealed in the topology of algebras or its structure as an infinite-dimensional vector space. These theories are usually studied in the context of real and complex numbers and functions. Analysis evolved from calculus, which involves the elementary concepts and techniques of analysis. Analysis may be distinguished from geometry; however, it can be applied to any space of mathematical objects that has a definition of nearness (a topological space) or specific distances between objects (a metric space). Mathematical analysis formally developed in the 17th century during the Scientific Revolution, but many of its ideas can be traced back to earlier mathematicians. Early results in analysis were implicitly present in the early days of ancient Greek mathematics. The book mathematical analysis emphasizes on metric spaces, convergence of sequences and series, continuity, differentiability, Riemann integral, sequences and series of functions, uniformity, interchange of limit operations.

CONTENTS

1

ON GEOMETRICAL METHODS THAT PROVIDE A SHORT PROOF OF FOUR COLOR THEOREM

Bahman Mashood*

* La Playa Street. 304. San Francisco.CA 94109, US

ABSTRACT

In this article we introduce a short and comprehensive proof of four color theorem based on geometrical methods. At the end of the article we will provide a short proof of the De Bruijn Erdos theorem for locally finite infinite graphs.

INTRODUCTION AND PRELIMINARIES

The Four Color Theorem is well known in the mathematical community. It states that any number of connected locally connected regions located in the plane (i.e' they are all subsets of R2), intersecting onlyw on their common boundaries, can be separated by four colors. In order to present a coherent mathematical view of the above theorem we need to state the following definitions and use the basic concepts and facts in topology. For more details on the basic concepts in topology we recommend the book, Topology (1970) by James Dugundji. Throughout this article the paths and closed curves are subsets of finite a graphs hence we use a type of digital geometry to prove some of the arguments. But many of the lemmas and theorems are stated and proved in the most general cases. The Jordan closed curve theorem as stated at [2] is the basis for many of the lemmas and theorems. When dealing with closed curves and paths we define the interior path of the given closed curve to be paths that except one or two points on the boundary the rest of the path lives in the interior of the closed curve. For a connected Graph G, its spectrum, spect(G) is the operator norm of the adjacency matrix. We say that two connected graphs G1 and G2 are equivalent, G1 ~ G2, if their corresponding adjacency matrix are unitary equivalent. this concept is used to define some useful definitions that will help proving the main theorem, theorem (2.21). the paths are all directed in same way. For example the paths located on closed curves have the same direction as the direction on the closed curve which is clockwise direction. Many of the sets we are going to deal with are connected, closed and bounded. These kinds of sets constitute a very important class of sets.

Definition 1.1 A compact connected subspace of a Hausdorf metric space is called continuum.

In particular here by region we mean locally connected continuum sub-set of R2. To state equivalent version of four color theorem represent each of the regions by a vertex in R2 and connect two points if and only if the corresponding regions intersect. This will provide us with a graph G. Now the four color theorem can be expressed equivalently to this situation as to associate a color to each of the vertices in such a way that no two vertices connected by an arc can have the same color. Note that without loss of generality we can assume that the above graphs are connected. In fact throughout this article all the graphs are connected. For a set $Q \subset R^2$, let Qc be the complement of Q in R^2, ie, $Q^c = (x \in R^2$, x not in Q). Also we denote the interior of Q by int(Q) In final part of this article we are going to deal with locally finite infinite graphs. These are infinite graphs with the property that each vertex is connected to finitely many other vertices. For a set Q, we set Card(Q) to be equal to the cardinality of the set Q. So by our definition any two region A2 and A2 from the collection are either disjoint or $(A1) \cap (A2) =$ (Bound $(A1) \cap$ Bound $(A2) \neq \phi$ In the later case A1 and A2 have to take different colors. Finally by Jordan closed curve theorem S divides the space in two regions, where one of the regions $SC \supseteq S$, S can be be mapped homeomorphically onto the closed disc corresponding to a circle. For simplifying the notation we write int(S) instead of writing int(SC).

The following definitions are well known

Definition 1.2 A subset of $S \subseteq R^2$ which is homeomorphic to a cicle is called closed curve. The above homeomorphism can be chosen so that can be extended to the homeomorphism taking SC = SUint(S) onto the closed disc that corresponds to the above circle.

Similarly the the edge or the arc connecting two vertices of a graph is homeomorphic to a closed interval. A path is a sub continuum of R^2 which is the union of number of arcs. It is clear that the path is also homeomorphic to a closed interval. Suppose S is a closed curve which is the subset of a connected graph G. Let us adopt the following notations.

Definition 1.3 Given a connected graph G, we denote the set of its vertices by V (G)), the number of its vertices by n(G), and the set of its edges by E(G). Suppose $S \subseteq G$ be a closed curve. We set $SCG = SC \cap G$ and SCintG, to be equal to the set SCG minus E(G) ie' $SCintG = SCG \cap (E(G))^c$.

Note that the set SCintG is not necessarily a connected set. Let G be a connected graph and suppose $(a_i)_{i=1}^{i=n(G)} \subset G$ be the set of all vertices of G. Furthermore for the vertices located on a closed curve S $\subseteq$ G the sequence $(a_i) \subseteq S$ of vertices have clockwise orientation, ie' a_{i+1} is located to the right of a_i. a_i and a_{i+1} are called neighbouring vertices if they are connected by an edge in V(G).

Definition 1.4 For a given connected graph G, The function f from V(G) to the set (1,2,3,4) is called colorization if for any two neighboring vertices a_i and , a_{i+1}, $f(a_i) \neq f(a_{i+1})$.

Definition 1.5 Following the same notations as in the above we define Cl(G), to be the set of all colorizations for G.

For any two points x,y $\in$ G,, let d(x,y) be the usual Euclidean distance of x from y in R^2. Let us set the following notations, d(x,G) = min (d(x,y), y $\in$ G) and d(G) = max (d(x,y), x, y$\in$G). For each x $\in$ G let D(x,d(x,G)) be equal to the closed Disk with the center in x and radius d(x,G). Let us define $D(G) = \cap_{x \in G}(D(x, d(x,G)))$. Then it is easy to check that D(G) is a closed set containing G.

Definition 1.6 Keeping the same notations as in the above, the vertex x $\in$ G is called to belong to the exterior boundary of G, denoted by EBound(G), if x can be connected via a path in $R^2 \cap (G)^c$ to any point in D(G)c $=R^2$ - D(G) = ((y $\in R^2$), y not in D(G)).

In the following arguments in order to shorten the notations for a given closed curve S we writeint(S) instead of int(SC).

Definition 1.7 Keeping the same notations as in the above, The closed curve $S \subseteq G$ is called maximal closed curve if and only if $S \subseteq$E Bound(G). It is called max closed curve if given some other closed curve S_1, with int(S_1) $\supseteq$ int(S), then S_1 = S.

Suppose S is a closed curve, $S \subseteq G$. Now let $x_1, x_2, \ldots x_m$ be number of vertices on $S \cap G$, where xi can be reached from x_{i-1} moving clockwise on encounter once we move from xi to x_{i+1} including the two end points. By $G(x_i, x_{i+1})$ we mean the set of all the vertices of $G \cap S$ that we encounter once we move from x_i to x_{i+1} excluding the two end points. Note that since we are moving clockwise on the closed curve we have the following identifications, $x_{m+1} = x_1, x_{m+2} = x_2, \ldots\ldots$ and so on. Note that moving clock wise along S, the intervals $[x_i, x_j]$, (respectively (x_i, y_i)) are well defined only if $|i-j| \leq m-1$. Suppose $a_1, a_2, \ldots a_m$ be the set of all the vertices on S moving clockwise on S. Then for each integer $1 \leq i$, the vertices a_i and a_{i+1} are called neighbouring vertices. Note that S is a metric space with the metric inherited from R^2. Given two points $x_1, x_2 \subseteq S$ where x_2 is to the right side of x_1 (ie' moving clockwise from x_1 we reach x_2), we set an order $x_1 \leq x_2$. the closed and open intervals on the closed curve S have the same definition as the ones on real line, just with the points belong to S, ie, $[x,y] = (z \in S)$, $x \leq z \leq y$, and $(x,y) = (z \in S)$, $x < z < y$. In order to identify the intervals on S clearly, we let the set $(a_i) \subseteq S$ to take their indices from the well known ring Z_m. The following theorem is going to be used throughout the rest of the article.

Theorem 1.8 Keeping the same notations as in the above, two max closed curves that are subsets of G can not intersect in more than one point,

Proof Suppose S_1 and S_2 are are max closed curves that are subset of G. Let us give clockwise orientation to the above closed curves. Since S_1 and S_2 are not equal there exits a point x in S_1 not belonging to S_2. Moving from point x along S_1 clockwise suppose S_1 intersect S_2 at vertices y_1 and y_2. With y_1 the first vertex on S_2 that S_1 will encounter and y_2 the last one. Now consider the paths, $[x,y_1] \subseteq S_1$ $[y_1,y_2] \subseteq S_2$, and $[y_2,x] \subseteq S_1$. It is clear that S_3 the union of these three, paths is a closed curve. In order to prove the theorem it is enough to show the following condition, $int(S_3) \supseteq (int(S_1) \cup int(S_2))$ holds. We call the above condition the U-condition. Let $v \in D(G)^c$ and w be a point in $int(S_1) \cup int(S_2)$. For U-condition to hold it is enough to show that any path s connecting v to w will intersect S_3. To complete the proof we need to prove the following lemmas.

Lemma 1.8.1 Keeping the above notations suppose $S_1 \supseteq [y_1,y_2] = s_1 = (S_1) \cap (S_2)$ then the U-condition holds. Futhermore S_3C contains s_1.

Proof Let $x \in int(S_1)$ and $v \in D(G)^c$. Suppose s is a path connecting v to x. Then s has to intersect the boundary of S_1. Hence s either hits $[y_2,y_1] \subseteq S_1$ or it hits $[y_1,y_2] \subseteq S_1$. Suppose in its way to v it hits $y \in [y_1,y_2] \subseteq S_1$ for the last time at point $y \in [y_1,y_2] \subseteq S_1$. Now using the fact that $int(S_1) \cap int(S_2) = \phi$, at point y, s either enters the interior of S_1 or enters the interior of S_2. In the first case it will hit $[y_2,y_1] \subseteq S_1$ before reaching v. In the second case it will hit $[y_1,y_2] \subseteq S_2$ before reaching v. This implies that $int(S_1) \cup int(S_2) \subseteq int(S_3)$ thus the U-condition hold. Now we clam that $int(S_3) \supseteq (y_1,y_2) \subseteq S_1$. To show that suppose $x \in (y_1,y_2) \subseteq S_1$ Then for $y \in int(S_1)$ there exists a path $s_2 = [x,y] \in int(S_1)$. But if x is in $(S_3C)^c$, then s_2 must intersect S_3. At this point the fact that S_3 does not intersects $(y_1,y_2) \subseteq S_1$ implies that $int(S_3) \supseteq (y_1,y_2) \subseteq S_1$.

Definition 1.8.2 The encounter of the closed curves Si, i=1,2, as stated in the above lemma is called of type-I

Lemma 1.8.3 Keeping the above notations suppose that S_1 and S2 intersect only on two points y_1 and y_2. Furthermore assume that moving clockwise from x on S_1 we enter the interior of S2 at y_1 and leave it at y_2. Then the U-condition holds.

Proof Note, it is easy to see that moving clockwise on S2 it will enter the interior of S_1 at y_2 and leave it at y_1. Let us consider the following paths. $p_1 = ([y_1,y_2] \subseteq S_1)$ and $p_2 = ([y_2,y_1] \subseteq S_2)$. Next define the closed curve $S_4 = p_1 \cup p_2$. Considering S_1 and S4, note that $S_1 \cap S_4 = p_1$. This using lemma 1.8.1 implies that $int(S_3) \supseteq int(S_1)$. Similarly we can show that $int(S_3) \supseteq int(S_2)$, and this complete the proof.

Definition 1.8.4 The encounter of the closed curves Si, i= 1,2, as stated in the above lemma is called of type-II.

Lemma 1.8.5 Keeping the above notations suppose that S_1 and S2 intersect only on two points y_1 and y_2. Furthermore suppose $S_1 \cap int(S_2) = S_1 \cap int(S_2) = \phi$. Then the U-condition holds

Proof Let us define the closed curves S4 and S5, S4 being equal to the union of $[y_2,y_1]\subseteq S_2$ and $[y_1,y_2]\subseteq S_1$. S5 being equal to the union of $[y_1,y_2]\subseteq S_2$ and $[y_1,y_2]\subseteq S_1$. It is easy to see that $int(S_2)\cap int(S_4)=\phi$, hence by the above lemma $int(S_5)\supseteq int(S_4)\cup int(S_2)\cup((y_2,y_1)\subseteq S_2)$ Where the directions of paths on Si, i= 1,2 are the clockwise directions on Si. Since $S_1\cap S_5=S_1=([y_1,y_2]\subseteq S_1)$ by lemma 1.8.1, $int(S_3)\supseteq int(S_1)\cup int(S_2)$ and this complete the proof

Definition 1.8.6 The encounter of the closed curves Si, i=1,2, as stated in the above lemma is called of type-III

In general following the same procedure as in the above moving clockwise on S_1 from x, suppose S_1 intersect S2 first time at y_1 and last time at y_2. Then let us define the closed curve S3, to be equal to the union of $[y_2,y_1]\subseteq S_1$ and $[y_1,y_2]\subseteq S_2$ Now set $S_{2,1}$ to be equal to the union of $[y_1,y_2]\subseteq S_1$ and $[y_1,y_2]\subseteq S_2$. But $S_1\cap S_{2,1}=[y_2,y_1]\subseteq S_1$. Thus since $int(S_1)\cap int(S_{2,1})=\phi$, the encounter of S_1 and $S_{2,1}$ is of type-I. Hence by lemma 1.8.1 $int(S_3)\supseteq int(S_1)$, similarly we can show that $int(S_3)\supseteq int(S_2)$ and this complete the proof of theorem 1.8.

Lemma 1.9 Suppose S_1 is a closed curved which is the subset of a connected graph G. Then there exist a max closed curve $S\subseteq G$, such that $int(S)$ contains $int(S_1)$

Proof For a given closed curve $S_2\subseteq G$ if $int(S_2)\supseteq(S_1)$ then we replace S_1 by S2. Next Using lemma 1.8 if S_1 intersect another closed curve $S_2\subseteq G$ in more than two points, then by the above arguments there exits another closed curve $S_3\subseteq G$ such that $int(S_3)\supseteq int(S_1)\cup int(S_2)$. Now proceed inductively, after finite number of steps we will end up having a closed curve $S\subseteq G$, such that if S intersects any other closed curve $S_k\subseteq G$ at more than one point then $int(S)\supseteq int(S_k)$. Further more if $S_l\subseteq G$ is any other closed curves with $int(S)\subseteq int(S_l)$, then $S=S_l$.

Theorem 1.10 Let G be a connected graph. Then the exist a set of max closed curves $(S_i)_{i=1}^{i=m}$, and paths $(s_i)_{i=1}^{i=m}$, such that G is equal to the union of the above paths and the set Q, $Q=\bigcup_{i=1}^{i=m}(S_i C\cap G)$. Furthermore $\text{EBound}(G)=\left(\bigcup_i(S_i)_{i=1}^{i=m}\right)\cup\left(\bigcup_i(s_j)_{j=1}^{j=m}\right)$

Proof In order to prove the above theorem we need the following definition and lemma

Let us consider a max closed curve $S_1\subseteq G$, with G a connected graph. Suppose we have a sequence of max closed curves $S_i\subseteq G$, $i=1,2,\dots,n$. Suppose for each $i<n$, S_i is connected to S_{i+1}, via a path in G.

Definition 1.11 Keeping the same notation as in the above. The sequence of closed curves (S_i), i = 2,3, …, n are called descenders of S1.

Lemma 1.12 Keeping the same notation as in the above, Let $S_i\subseteq G$, i =1.2.3 ,.. be a sequence of max closed curves. Then the above max curves can be connected to each other only by unique path $s\subseteq G$ Furthermore if S2 and S3 are both connected to S1 via paths in G, then S2 and S3 or any of their descenders can not be connected to each other via paths in G.

Proof Follows from the definiton of max closed curves and the arguments of lemma 1.8.

Note that the above statements holds if we consider the vertices of the paths connecting the max closed curves as max closed curves. In particular if we retract each one of the max closed curves to a point the resulting will be a tree.

Finally theorem 3.22 implies that any one of max closed curve or corresponding connecting paths between them are subset of EBound(G). But by theorem(1.10) any vertex of $z \in G$ is either a member of $S_i C \bigcap G$ for some max closed curve $S_i \subseteq G$ or is a vertex on one of the paths $s_{i,j}$ connecting two max closed curves $S_i \subseteq G$ and $S_j \subseteq G$. This complete the proof of theorem 1.10.

The following lemmas can be verified immediately therefore we skip the proofs.

Lemma 1.13 Keeping the same notations as before. Given a path $s \subset G$, then the vertices on s can be separated by using two colors.

Lemma 1.14 Given a closed curve $S \subset G$, with G a connected graph. Then if the number of vertices on S, is even then the vertices on S can be separated by two colors. Otherwise three color would be sufficient to separate all the vertices on S.

At this point suppose That G is a connected graph and $S \subseteq G$ be a max closed curve. Now let $S_1 \subseteq G$ be another closed curve with $S_1 C \subseteq SC$ and such that $S_1 C \operatorname{int} G = \phi$. Then G is equal to the union of $(S_1 C)^c$ and S1. Assuming that G is located on the surface of an sphere implies that there exists another connected graph G1 ~ G such that S1 is max closed curve in G1. We denote S1 a semi max close curve.

Definition 1.15 We say a connected graph G has property R. if there exist a colorization $cl \in Cl(G)$such that on every semi max closed curve $S \subseteq G$, $Card(cl(S)) \leq 3$.

Now it is clear to see that if G1 is a connected graph with property R and G2 another connected graph such that G1 ~ G2, then G2 has property R too. Next let G be a connected graph and $S \subseteq G$ a closed curve. Let (a_i) $i = 1, 2, ..., m$ be a sequence of all the vertices on S.

Definition 1.16 Let $a_i, a_j \in V(S)$. Given a path $s \subseteq SC \operatorname{int} G$ connecting a_i to b_j. Now consider the closed curve $S_1 \subseteq G$ that is the union of $([a_i, a_j] \subseteq S)$ and s. We call the path s reducible if there exists another internal path $s_1 \subseteq SC \operatorname{int} G$, between two different points $c \leq d \in G[a,b] \subseteq S$, such that for the closed curve $S_2 = ([c,d] \subseteq S) \bigcup s_1$, $S_2 C$ is a proper subset of $S_1 C$. Otherwise s is called irreducible. suppose there exist two vertices $a_i, a_j \in V(S)$ and irreducible path in $int(S)$, connecting a_i to a_j. If $j - i > 1$ we say S is of type-I. Otherwise if j = i + 1, we say S is of type-II. If S is not of type-I or type-II, we say S is of type-III.

Lemma 1.17 Given a closed curve $S \subseteq G$ with G a connected graph. Let (a_i), $i = 1, 2, ..., m$ be the set of vertices of S moving clockwise on S. Suppose there exists a vertex $a_j \subset S$, such that the connected component of SCG that contains a_j and does not contains any edges in S, contains another vertex $S \ni a_k \neq a_j$. then there exits an irreducible path in $SC \operatorname{int} G$ connecting a_j to a vertex $a_j \neq a_l \subseteq S$.

Proof If a connected component $G_1 \subseteq G$ as in the above contains a_j and another vertex $a_k \neq a_j$, then by the fact that G_1 is connected graph too, we have a path $s \subseteq G_1$ that connect a_j to a_k. Then it is clear that there exits a vertex $a_l \in S$ and irreducible path $s \subseteq G$, connecting a_j to a vertex $a_j \neq a_l \in S$.

At this point we are going to introduce an special subset $\Gamma \subset R^2$, $\Gamma = (n(G), Spect(G))$, G a connected graph. We impose ordering $\leq$ on Γ by $\Gamma \ni (x_1, y_1) \leq (x, y_2) \in \Gamma$ if $x_1 \leq x_2$ and $y_1 \leq y_2$ It is easy to see that $\leq$ is partial ordering on Γ. Now using the fact that for every integer k the set spect(k) = (spect(G), G a connected graph and n(G) = k) is a finite set, we can order the above set by the magnitude of its members. Hence set $spect(k) = (r_i)_{i=1}^{i=m_k}$. Where for i > j, $r_i > r_j$, r_1 = lower bound(spect(k)) r_{m_k} = upper bound(spect(k). Also to facilitate our notations for a closed curve $s \subseteq G$ we replace Cl(SCG) by Cl(S).

Lemma 1.18 Let S_1 and S2 be subsets of connected graph G both being max closed curves. Suppose the above two closed curves are connected by a path $s \subseteq G$ or a common vertex q and that they have colorizations $cl_i \in Cl(S_i)$, i = 1,2. Let us define connected graph $G_1 = S_1 \cup S_2 \cup s$. Then There exists a colorization $cl_i \in Cl(G_1)$ extending cl_1 or cl_2.

Proof By lemma 1.12 Points on s can be separated by two colors, Thus cl_1 can be extended to $S_1 \cup s$ in an obvious way. Suppose the point $b \in S_2 \cup s$ is connected by an edge in s to the point c 6= b in s. If cl1(b) = i = cl2(b) = j then we are done. Otherwise if i ≠ j we define cl3 ∈ Cl(S2) by cl3(a) = i, if cl2(a) = j, cl3(a) = j, if cl2(a) = i or else cl3(a) = cl2(a). Now cl1 can be extended to G1 and we are done in this case. Finally If S_1 and S2 are connected by a common vertex q and cl1(q) = i = cl2(q) = j then we are done. Otherwise repeating the above arguments we can define a colorization cl3 on S2 U S_1 = G1 and we are done.

At this point note that every connected subset the connected graph G is also a connected graph. As before we define an interior graph to be a subgraph of G that does not contain any element from the set E(G).

Theorem 1.19 Given a connected graph G Suppose every max close curve, $S \subseteq G$ has property R. Then G has a colorization.

Proof Suppose G is a connected graph and Let $(S_i)_{i=1}^{i=m}$ be the set of all max closed curves in G. Let $(s_j)_{j=1}^{j=l}$ be the set of all paths in G connecting max closed curves. Where some of these max closed curves consists of single vertex. Therefore by theorem 1.10 and lemma 1.12 we can assume that G is equal to the union of the max closed curves, the part of G they contain in their interior and the paths in G that connecting them. By lemma 1.12 there are unique pathes between the maximal closed curves and for any two maximal closed curve that are connected to a third one their descenders are not connected to each others via paths or common vertices. Now our assumption states that for any one of max closed curve $S \subseteq G$, SCG has property R. Thus each of the maximal closed curves $S_j \subseteq G$ has a colorization clj acting on S_jCG. Next let us begin from the max closed curve S1 with a colorization $cl_1 \in S_1C \subseteq G$ and all max closed curves $S_{1,r} \subseteq G$ that are connected to S1 by the paths $S_{1,r} \subseteq G$. For each of the max closed curves S1,r, set $G_{1,r} = S_{1,r}CG \cup s_{1,r}$. Now by lemma 1.12, we can extend cl1 to become a colorization on $G_{2,r} = S_1CG \cup G_{1,r}$. Now repeating the above process from each of the max closed curves S1,r, using lemma 1.18 we are going to extend cl1to become a colorization on each of the descenders of S1 and ultimately after repeating the above process finitely many times we will get a colorization acting on G.

THE MAIN RESULT

In this section we are going to state the main theorem and its proof.

Lemma 2.20 Suppose G is a connected graph and S a subset of G which is the max closed curve with SCG = G. Then S has property R.

Proof In order to prove the theorem we need to state and proof the following lemmas

Lemma 2.20.1 Keeping the same notations as in the above suppose S is of type-II. Then there are two vertices $a_i \in G(S)$, $i = 1, 2$ and an irreducible path $s_1 \subseteq int(S)$ connecting the above two vertices. Furthermore there exist two connected graphs G1 and and G1 each containing a max closed curve, $S_1 \subseteq G_1$, $S^1 \subseteq G^1$, such that $S^1CG^1 = G^1$ and $S^1CG^1 = G^1$. Finally we have $int(S) \supseteq int(int(S^1)\ int(S_1) \supseteq int(int(S^1)$ and $G \sim G_1$.

Proof Set $S^1 = s_1 \cup (|a_1, a_2|) \subseteq S$, then $S^1 \subseteq G$ is a closed curve. Next we can construct a path $s_1 \subseteq G^c$ connecting a1 to a2. Let $G^1 = S^1CG$, $G_1 = G^1 \cup s_2$ and $S_1 = s_1 \cup s_2$. It is clear that $S_1 \subseteq G_1$ is a max closed curve in G1 and $S_1CG_1 = G_1$. Next following their constructions, $int(S_1) \supseteq int(S^1)$ and $int(S) \supseteq int(S^1)$.

Before proceeding to the next lemma we need to bring the following definition

Definition 2.20.2 Let G1 and G2 be a connected graph and $S_i \subseteq G_i$, $i = 1,2$ be a max closed curves with SiCG = Gi, i=1,2. then we call Si a full closed curve. In particular if G1 ~ G2 then we write S1 ~ S2.

Lemma 2.20.3 Let $S \subseteq G$ be as in the above. Then S is either equivalent to type-I or to type-III closed curve.

Proof If S is type-I or type-III we are done. Otherwise following notations and arguments of lemma 2.20.1, consider the full closed curve $S_1 \subseteq G_1$ and $S^1 \subseteq G$. G1 connected graph. We had S ~S1, $int(S_1) \supseteq int(S^1)$, $V(S1) = V(s_1) = V(G) \cap s_1$, $int(S) \supseteq int(S^1)$. Furthermore any path in the interior of S1 connecting to vertices of S1 lives in the interior of S1. Now if S1 is of type-I or type-III we are done, otherwise proceeding as in the above we have two closed curve S2 and $S^2 \subseteq G$with the following properties. S2 is a full max closed curve subset of a connected Graph $G2 \sim G$, $int(S_2) \supseteq int(S^2)$, $int(S_1) \supseteq int(S^2)$, $int(S^1) \supseteq int(S^2)$ and the interior paths connecting vertices of S2 live in the interior of S2. Proceeding by induction and because G is a finite graph after finitely many steps we get the sequence of closed curves, Si and Si $1 \leq I \leq m0$ with the following properties. Si ~S, $int(S_i) \supseteq int(S^i)$,

$int(S^{i-1}) \supseteq int(S^i)$ and such that any interior paths connecting the vertices of Si lives in the interior of Si. Hence by the fact that G is finite at some stage m0, either G_{m0} is of type-I or of type-III. The proof of the lemma is complete.

To complete the proof of the lemma we use induction on n(G). So suppose given an integer k such that the statement of lemma hold for all connected graph having only one max closed curve S with $n(G) \leq k$. Lets a1,a2, …am be a set of all vertices on $S \subseteq G$, where ai can be reached from ai−1moving clockwise on S. If all connected component of G intersect S at most at one point, ie, S is of type-III. In this case by lemma 1.14 we can construct a colorization $cl \in Cl(S)$, $cl(S) \leq 3$. Now we want to extend cl to become a colorization on all G. Since for each of the connected components $G_i \subseteq G$, $n(Gi) \leq k$, then by induction and theorem 1.19, Gi has property R. Now using lemma 1.18 we can extend cl to become a colorization for G thus G has property R. Otherwise by the above lemma 2.20.3 we can assume that S is of type-I. Hence for some vertex aj 2 S there exits an irreducible interior path $s \subseteq$ SCintG from aj to al, where aj and al vertices of S, moving clockwise on S with j + 1 <l. Therefore we will get a closed curve $S_1 \subseteq G$ which is the union of s and the path $[al,aj] \subseteq S$. Now note that S1CG is a connected graph with n(G1 = S_1 $C \cap G$) = k.. Hence by induction assumption G1 = S1CG has property R therefore there exists a colorization $cl \in Cl(G_1)$ such that $Card(cl(S1)) \leq 3$. Now note that any connected component of G that contains a vertex in $G[aj,al] \subseteq S$, does not intersect S at any other point. Hence using lemma 1.14 and lemma 1.18 we can extend cl to become a colorization on G with $Card(cl(S)) \leq 3$. This complete the proof of lemma 2.20.

Theorem 2.21 every connected graph G, has property R thus has a colorization.

Proof By lemma 2.20 each one of max closed curves has property R. Finally Theorem 1.19 completes the proof.

Finally we are going to bring a short proof of De Bruijn Erdos theorem for locally finite infinite planar graphs

Theorem(De Bruijn Erdos) Suppose G is a locally finite infinite graph. Then G has a colorization.

Proof Note that without loss of generality we can assume that G is connected. Let $v \in G$ be a vertex in G. Consider all the paths $s \in E(G)$ of length $m \in N$ from v to other vertices in G. The union of these paths is a finite connected subgraph $G_m \subseteq G$. By theorem(2.21), there exists a colorization $cl_m \in Cl(Gm)$. Next Cl(Gm) can be considered as a subset of $R^{n(G_m)}$ with integer entries. Let us pick randomly one of these graphs say $G_1 = G_{m0}$. Thus there exist an infinite subsequence ($m \leq$ mi)i of (N), and a colorization $cl_{1,\infty} \in Cl(G^1)$ such that for each Gmi the restriction of clmi to G1 will agree with cl1,∞. Proceeding inductively we can construct an increasing sequence of finite subgraphs of G, $G^1 \subseteq G^2 \subseteq \ldots\ldots \subseteq G^k \subseteq \ldots$, where for each $\leq k$, Gk has a colorization clk,∞, with the property that the restriction of clk+1,∞ to Gk will be equal to clk,∞. But G =S(Gk)k, and this complete the proof.

APPENDIX

In this section we are going to proof some technical lemmas needed in the proof of number of the lemmas and theorems in the above

Theorem 3.22 Keeping the same notations as in the above. let $S \subseteq G$ be a max closed curve. Then S *EBound(G)*.

Proof As we demostrated, if (Si), i = 1,2, … m be the set of all max closed curves, then G is the union of the sets SiCG together with the paths si,j, i,j = 1, 2, …, m, where si,j is the unique path connecting Si to Sj. By the finiteness of the Graph G we can assume that each of the max closed curves is a circle and the connecting paths are straight lines going through the centers of corresponding circles. Next for each of the circles Si there exists a circle Si,l with the same center but strictly larger than Si. By taking Si,l close enough to Si we can make sure that Si,l does not intersect any other max closed curves or the paths connecting them to any other max closed curve other than Si. set bi,j be the intersection of si,j with Si,l. Now moving clockwise on Si,l, consider two points bi,j,1 < bi,j < bi,j,2 on Si,l. We can choose them close enough so that if we draw lines si,j,1 and si,j,2, parallel to si,j intersecting Sj,l at bi,j,2 and bi,j,1 respectively, with bj,i,2 > bj,i,1, then si,j,k, k = 1, 2 intersect G at Si and Sj only. Next let aj,i,k, k = 1, 2, be the intersection of si,j,k, k = 1, 2 with Sj respectively. let us define the closed curve Si,j,s to be defined to be the union of $[a_{i,j,1}, a_{j,i,2}] \subseteq$ si,j,1, $[a_{j,i,2}, a_{j,i,1}] \subseteq S_j$, $[a_{j,i,1}, a_{i,j,2}] \subseteq s_{i,j,2}$ and $[a_{i,j,2}, a_{i,j,1}] \subseteq S_i$. Then it is clear that Si,j,s contains si,j in its interior. At this point we assume that Si is connected to the sequence of max closed curves (Sj), j = 2, …, m1, where each of the above closed curves is connected to Si only. Next for each integer 2 ≤ j ≤ m we define a loop Ω(i,j) ⊆ (G)c to be a path which is the union of following paths, $\left[b_{i,j,1}, b_{j,i,2}\right] \subseteq s_{i,j,1}, \left[b_{j,i,2}, b_{j,i,1}\right] \subseteq S_j$ and $\left[b_{j,i,1}, b_{i,j,2}\right] \subseteq s_{i,j,2}$. Now suppose Sj is connected to more than one closed curve $S_k \subseteq G,\ k = 2,..m_{1,j}$ Ordering the points of intersection of sj,k with Sj. Let aj,k, $j = 1,2,..,m_1$ be the sequence of the above vertices. At this point we assume that The sequence of closed curves that are connected to Sj, j = 1,2, …, m1 are not connected to any other max closed curve. Next for a fixed 2 ≤ j1 ≤ m1 and $2 \le k_1 \le m_{1,j_1}$ we defined the loop $\Omega(j_1,k_1) \subseteq G^c$ to be a path going from $b_{j_1,k_1,1} \in S_{j_1,l}$ to $S_{k_1,l}$ and back to $S_{j_1,l}$. Now we want to extend the definition of loop to more complicated case. we want to define a path $\Omega(i, j_1) \subseteq G^c$., going from $b_{i,j_1,1} \in S_{i,l}$ to $S_{j_1,l}$ and back to Si,l. Set $\Omega(i, j_1)$ to be equal to the union of the following paths, $\left[b_{i,j_1,1}, b_{j_1,i,2}\right] \subseteq s_{i,j_1,1}$, $\left[b_{j_1,i,2}, b_{j_1,i,1}\right] \subseteq S_{j_1}$, $\left[b_{j_1,i,1}, b_{i,j_1,2}\right] \subseteq s_{i,j_1,2}$ and $\bigcup_{k=2}^{k=m_{1,j_1}} \left(\Omega(j_1,k) \cup \left[b_{j_1,k,2}, b_{j_1,(k+1),1}\right]\right)$. Using our assumptions each one of the paths $\Omega(j_1,k)$., is well defined hence the above formula is going to identify the path $\Omega(i, j_1) \subseteq G^c$. We call the above formula the loop formula. We saw that the loop formula is valid for two special cases. We want to show that the loop formula will identify the path $\Omega(i, j_1) \subseteq G^c$ from $S_{i,l}$ to $S_{j_1,l}$ and back to $S_{i,l}$ in the most general case. By the loop formula we can identify $\Omega(1, j_1) \subseteq G^c$, if we can identify $\Omega(j_1, k)$ for all the max closed curves connected to S_{j_1} except Si. Similarly to identify Ω(j1, k), it is enough to identify all the loops Ω(j2, k) where $S_{j_2} \neq$ Si is one of the max closed curves that are connected to S_{j_1}, furthermore for each $2 \leq$ k, Sk is one of the max closed curves that is connected to $S_{j_2} \neq S_{j_1}$. We call this the stage (j1, j2) of calculation. inductively we have stages (j1, j2, …, jn) of calculation. If there exists some integer n0 such that for n $\geq$ n0 at every stage (j1, j2, …, jn) we can calculate (jn, k) for all max closed curves that are connected to S_{j_n} and that are different from $S_{j_{n-1}}$, then we are done. But because of the finiteness of G, there exists an integer n0 such that for n $\geq$ n0, Sjn is only connected to $S_{j_{n-1}}$. This implies that at the stages (j1, j2, …, j_{n_0-1}) we can calculate (j_{n_0-1},k), so by induction we are done. This implies the following lemma

Lemma 3.22.1 Keeping the same notations as in the above. For every connected max closed curves Si and Sj there exists a path Ω(i, j) $\subseteq G^c$ going from Si,l to Sj,l and back to Si,1.

To complete the proof of the theorem we have to show that for any point x $\in$ G, and v $\in D(G)^c$, there exists a path $s \subseteq G^c$ connecting x to v. Next We need to employ some new notation. As we mentioned we say two max closed curves $S_1 \subseteq G$ and $S_2 \subseteq G$ are connected, if they are connected by the path s1,2. By chain of max closed curves we mean a sequence (Si), i = 1, 2, …, m such that for each i $\leq$ m, Si and Si+1 are connected. Now without loss of generality we can assume $S_1 \ni x = a_{1,j_1,1}$ for some max closed curve $S_1 \subseteq G$, where S_{j_1}, is a closed curves connected to S1. The other cases can be overcome using some minor technicalities. Now there exists a path s1 $\subseteq (S_1)^c$ from a_{1,j_1} to $b_{1,j_1,1}$, furthermore by the structure of G we can assume that in fact $s_1 \subseteq G^c$. Also there is a path $s_2 \in S_{1,l}^c$ from $b_{1,j_1,1}$ to v. Moving from v to $b_{1,j_1,1}$ on s2 suppose the first point that we intersect G is y = ak,k+1 $\in$ Sk, Where S_k, $S_{k+1} \subseteq G$ are connected max closed curves. The other cases can be overcome with minor technicality. At this point without loss of generality we can assume that s2 will intersects Sk,1 first time at $b_{k,l,1}$. But by the structure of G, the exists a chain of closed curves $\left(S_{i,1}\right)_{i=1}^{i=k}$ rom $S_{1,l}$ to $S_{k,l}$. Next let us define the path $s_3 \subseteq G^c$, by $s_3 = \bigcup_{i=j_1+1}^{i=2+j_1}\left(\Omega(1, i-1) \cup \left[b_{1,i-1,2}, b_{1,i,1}\right]\right)$. In this paths all the intervals are on S1,l and it takes point $b_{1,j_1,1}$ to point $b_{1,2,1}$. Note since we number the max closed curves connected to S_1 S1 by moving clockwise on S_1 by our definition, $S_2 = S_{2+j_1}$. Let us denote s_3 by $N\left(b_{1,j_1,1}, b_{1,2,1}\right)$. (It is also clear that if none of the points in the interval $\left[b_{1,j_1,1}, b_{1,2,1}\right] \subseteq S_{1,l}$ were connected to another max closed curve then $N\left(b_{1,j_1,1}, b_{1,2,1}\right) = \left[b_{1,j_1,1} b_{1,2,1}\right]$. Now for an integer i $\leq$ k, consider the path s(i, i + 2) $\subseteq G^c$, that is the union of $\left[b_{1,j_1,1}, b_{1,2,1}\right] \subseteq S_{1,l}$ and $N\left(b_{i+1,i,2}, b_{i+1,i+2,1}\right)$. Finally consider the paths $s_4 = \left(\bigcup_{i=2}^{i=k-2} s(i, i+2)\right) \subseteq G^c$, $s_5 = \left[b_{k,k+1,1}, v\right] \subseteq s \cap G^c$ and $s_6 = \left[a_{k,k+1}, b_{k,k+1}\right] \subseteq G^c$. At this end the path

$s_7 = s_5 \cup s_6 \cup s_4 \cup s_3 \cup s_1$ is a path in Gc taking x to v and this complete the proof of the theorem.

REFERENCES

1. Allaire, F, "Another proof of the four colour theorem Part I", Proceedings, 7th Manitoba Conference on Numerical Mathematics and Computing, Congr. Numer. 20: 372. (1997).
2. Appel, Kenneth; Haken, Wolfgang, "Every Planar Map is Four Colorable Part I. Discharging", Illinois Journal of Mathematics 21: 429490. (1977).
3. Appel, Kenneth; Haken, Wolfgang; Koch, John, "Every Planar Map is Four Colorable Part II. Reducibility", Illinois Journal of Mathematics 21: 491567. (1977).
4. Appel, Kenneth; Haken, Wolfgang, "Solution of the Four Color Map Problem", Scientific American 237 (4): 108121, (October 1977).
5. Appel, Kenneth; Haken, Wolfgang, Every Planar Map is Four-Colorable, Providence, RI: American Mathematical Society. (1989).
6. Bernhart, Frank R, "A digest of the four color theorem.", Journal of Graph Theory 1: 207225, (1977).
7. Borodin, O. V, "Solution of the Ringel problem on vertex-face coloring of planar graphs and coloring of 1-planar graphs", Metody Diskretnogo Analiza (41): 1226, 108, MR 832128. (1984).
8. Cayley, Arthur, "On the colourings of maps", Proc. Royal Geographical Society (Blackwell Publishing) 1 (4): 259261, (1879).
9. Fritsch, Rudolf; Fritsch, Gerda, The Four Color Theorem: History, Topological Foundations and Idea of Proof, New York: Springer, (1998).
10. Gonthier, Georges, "Formal Proof The Four-Color Theorem", Notices of the American Mathematical Society 55 (11): 13821393, (2008).
11. Gonthier, Georges, A computer-checked proof of the four colour theorem, unpublished
12. (2005) Hadwiger, Hugo, "ber eine Klassifikation der Streckenkom-plexe", Vierteljschr. Naturforsch. Ges. Zrich 88: 133143. (1943).
13. Hale, Thomas C, The Jordan Curve Theorem formaly and informaly, The American Monthly 114(10):889-894.
14. Wilson, Richard and Newmann Victort Lara, Digital Jordan curves, Topology and its applications, Volume 46, Issued 3, 30 October (1992).
15. Tymchatyn, Ed,(Department of mathematics University of Saskatchewan), exchange of ideas and communications. (2014).

2

SOLUTIONS OF LINEAR FRACTIONAL NON-HOMOGENEOUS DIFFERENTIAL EQUATIONS WITH JUMARIE FRACTIONAL DERIVATIVE AND EVALUATION OF PARTICULAR INTEGRALS

Uttam Ghosh[1, *], Susmita Sarkar[2], Shantanu Das[3,4]

[1]Department of Mathematics, Nabadwip Vidyasagar College, Nabadwip, Nadia, West Bengal, India

[2]Department of Applied Mathematics, University of Calcutta, Kolkata, India

[3]Reactor Control System Design Section Bhabha Atomic Research Centre, Mumbai, India

[4]Department of Physics, Jadavpur University Kolkata, West Bengal, India

ABSTRACT

In this paper we describe a method to solve the linear non-homogeneous fractional differential equations (FDE), composed with Jumarie type fractional derivative, and describe this method developed by us, to find out particular integrals, for several types of forcing functions. The solutions are obtained in terms of Mittag-Leffler functions, fractional sine and cosine functions. We have used our earlier developed method of finding solution to homogeneous FDE composed via Jumarie fractional derivative, and extended this to non-homogeneous FDE. We have demonstrated these developed methods with few examples of FDE, and also applied in fractional damped forced differential equation. The short cut rules, that are developed here in this paper to replace the operator Da or operator D2a as were used in classical calculus, gives ease in evaluating particular integrals. Therefore this method proposed by us is useful and advantageous as it is having conjugation with the classical methods of solving non-homogeneous linear differential equations, and also useful in understanding physical systems described by FDE.

INTRODUCTION

The fractional differential equations and its solutions arises in different branches of applied science, engineering, applied mathematics and biology [1-9]. The solutions of fractional difference equations are obtained by different methods which includes Exponential-Function Method [10], Homotopy Perturbation Method [11], Variation Iteration Method [12], Differential transform Method [13] and Fractional Sub-equation Method [14], Analytical Solutions in terms of Mittag-Leffler function [15]. In developing those methods the usually used fractional derivative is Riemann-Liouvellie (R-L) [6], Caputo derivative[6], Jumarie's left handed modification of R-L fractional derivative [16, 17]. In [15] we have developed an algorithm to solve the homogeneous fractional order differential equations in terms of Mittag-Leffler function and fractional sine and cosine functions. However, there are no standard methods to find solutions of non-homogeneous fractional differential equations. In this paper we describe a method to solve the fractional order non-homogeneous differential equations. Organizations of the paper are as follows; in section 2.0 we describe the different definitions of fractional derivatives and properties of Mittag-Leffler function. In section-3.0 we describe the solutions of $\alpha-$ order fractional differential equations. In section 4.0 the solutions of $2\alpha-$ order fractional differential equations is described, with several types of forcing functions. In section 5.0 this methods has been applied to solve both un-damped and damped fractional order forced oscillator equations. In this paper the fractional derivative operator ${}^{J}D^{u}$ will be of Jumarie type fractional derivative.

DEFINITION OF FRACTIONAL DERIVATIVES

The useful definitions of the fractional derivatives are the Grunwald-Letinikov (G-L) definition and Riemann-Liouville(R-L) definition [6] and Modified R-L-definitions [16, 17].

Grunwald-Letinikov definition

Let $f(t)$ be any function then the α -th order derivative $\alpha \in \mathbb{R}$ of $f(t)$ is defined by

$${}_aD_t^{\alpha}\left[f(t)\right] = \lim_{\substack{h\to 0\\ nh\to t-a}} h^{-\alpha} \sum_{r=0}^{n} \binom{\alpha}{r} f(t-rh)$$

$$= \lim_{\substack{h\to 0 \\ nh\to t-a}} h^{-\alpha} \sum_{r=0}^{n} \frac{\alpha!}{r!(\alpha-r)!} f(t-rh)$$

$$= \frac{1}{(-\alpha-1)!} \int_a^t (t-\tau)^{-\alpha-1} f(\tau)d\tau$$

$$= \frac{1}{\Gamma(-\alpha)} \int_a^t \frac{f(\tau)}{(t-\tau)^{\alpha+1}} d\tau$$

Where α is any arbitrary number real or complex; and the generalized binomial coefficients are described as follows [1, 16, 17]

$$\binom{\alpha}{r} = \frac{\alpha!}{r!(\alpha-r)!} = \frac{\Gamma(\alpha+1)}{\Gamma(r+1)\Gamma(\alpha-r+1)}$$

The above formula becomes fractional order integration if we replace α by $-\alpha$ which is Riemann fractional integration formula

$$\int_a^t f(\tau)d\tau^{\alpha} = {}_aI_t^{\alpha}\left[f(t)\right] = {}_aD_t^{-\alpha}\left[f(t)\right]$$

$$= \frac{1}{\Gamma(\alpha)} \int_a^t (t-\tau)^{\alpha-1} f(\tau)d\tau \qquad \alpha > 0$$

In above we have noted several notations used for fractional integration.

Riemann-Liouville fractional derivative definition

Let the function $f(t)$ is one time integrable then the integro-differential expression as following defines Riemann-Liouvelli fractional derivative [1, 6]

$$
{}_aD_t^{\alpha}\left[f(t)\right] = \frac{1}{\Gamma(n-\alpha)}\left(\frac{d}{dt}\right)^{n}\int_a^t (t-\tau)^{(n-\alpha)-1} f(\tau)d\tau \quad \alpha > 0
$$

Here the n is a positive integer number just greater than real number α The above expression is known as the Riemann-Liouville definition of fractional derivative [6] with $(n-1) < \alpha < n$

In the above definition fractional derivative of a constant is non-zero.

Modified Riemann-Liouville definition

To overcome the shortcoming fractional derivative of a constant, as non-zero, another modification of the definition of left R-L type fractional derivative of the function $f(x)$, in the interval $[a,b]$ was proposed by Jumarie [16] in the form described below

$$
{}_a^JD_x^{\alpha}\left[f(x)\right] = \begin{cases} \dfrac{1}{\Gamma(-\alpha)}\displaystyle\int_a^x (x-\tau)^{-\alpha-1} f(\tau)d\tau, \ \alpha < 0. \\ \dfrac{1}{\Gamma(1-\alpha)}\dfrac{d}{dx}\displaystyle\int_a^x (x-\tau)^{-\alpha}\left[f(\tau)-f(a)\right]d\tau, \\ \qquad 0 < \alpha < 1 \\ \left(f^{(\alpha-m)}(x)\right)^{(m)}, \quad m \le \alpha < m+1. \end{cases}
$$

Here we state that $f(x)=0$ for $x<a$ and $x>b$. However in this paper we will be using this left-Jumarie fractional derivative that is ${}_0^JD_t^{\alpha}\left[f(t)\right]$, for $0<\alpha<1$ and with condition $f(t)=0$ for all $t<0$. We will simplify the symbol and drop $a=0$ and differentiating variable t and simply write ${}^JD^{\alpha}[f(t)]$. Using the above definition Jumarie [16] proved the following

$$
(u(x)v(x))^{(\alpha)} = u^{(\alpha)}(x)v(x) + u(x)v^{(\alpha)}(x)
$$

We have recently modified the right R-L definition of fractional derivative of the function $f(x)$, in the interval $[a,b]$ in the following form [17],

$$ {}_{x}^{J}D_{b}^{\alpha}\left[f(x)\right]=\begin{cases}-\dfrac{1}{\Gamma(-\alpha)}\displaystyle\int_{x}^{b}(\tau-x)^{-\alpha-1}f(\tau)d\tau, & \alpha<0. \\ -\dfrac{1}{\Gamma(1-\alpha)}\dfrac{d}{dx}\displaystyle\int_{x}^{b}(\tau-x)^{-\alpha}\left[f(b)-f(\tau)\right]d\tau, & \\ \qquad 0<\alpha<1 & \\ \left(f^{(\alpha-m)}(x)\right)^{(m)}, & m\le\alpha<m+1.\end{cases} $$

Using both the modified definition we investigate the characteristics of the non-differentiable points of some continuous functions in [17]. The above defined all the derivatives are non-local type, and obtained solution to homogeneous FDE, with Jumarie derivative [15]. Subsequently we will be using ${}^{J}D^{u}$ as fractional derivative operator of Jumarie type, with start point $a=0$, and stating the function $f(x)=0$ for all $x<0$ in following sections.

The Mittag-Leffler Function

The Mittag-Leffler function was introduced by Gösta Mittag-Leffler [18] in 1903. The one-parameter Mittag-Leffler function is denoted by $E_{\alpha}(t^{\alpha})$ and defined by following series

$$ E_{\alpha}(z)\overset{\text{def}}{=}\sum_{k=0}^{\infty}\frac{z^{k}}{\Gamma(1+\alpha k)}, \qquad z\in\mathbb{C}, \qquad \operatorname{Re}(\alpha)>0 $$

Again from the Jumarie definition of fractional derivative we have ${}_{a}^{J}D_{x}^{\alpha}[C]=0$ we apply this property to get α order Jumarie Derivative of the Mittag-Leffler function $E_{\alpha}(ax^{\alpha})$ as follows

$$
\begin{aligned}
{}_0^J D_x^\alpha\left[E_\alpha(ax^\alpha)\right] &= {}^J D^\alpha\left[E_\alpha(ax^\alpha)\right] \\
&= {}^J D^\alpha\left[1+\frac{ax^\alpha}{\Gamma(1+\alpha)}+\frac{a^2x^{2\alpha}}{\Gamma(1+2\alpha)}+\frac{a^3x^{3\alpha}}{\Gamma(1+3\alpha)}+\ldots\infty\right] \\
&= a\left(1+\frac{ax^\alpha}{\Gamma(1+\alpha)}+\frac{a^2x^{2\alpha}}{\Gamma(1+2\alpha)}+\frac{a^3x^{3\alpha}}{\Gamma(1+3\alpha)}+\ldots\infty\right) \\
&= aE_\alpha(ax^\alpha)
\end{aligned}
$$

Therefore the fractional differential equation ${}_0^J D_x^\alpha[y] = ay$ has solution in the form $y = AE_\alpha(ax^\alpha)$, where A is an arbitrary constant.

Non-Homogeneous Fractional Differential Equations and Some Basic Solutions

The general format of the fractional linear differential equation is

$$
f\left({}^J D^\alpha\right)y = g(t^\alpha) \tag{2.1}
$$

Where $f\left({}^J D^\alpha\right)$ isa linear differential operator $0 < \alpha < 1$, of Jumarie type. The above differential equation is said to be linear non-homogeneous fractional differential equation when $g(t^\alpha) \neq 0$, otherwise it is homogeneous. Solution of the linear fractional differential equations (composed via Jumarie derivative) can be easily obtained in terms of Mittag-Leffler function and fractional sine and cosine functions [15].

The function $g(t^\alpha)$ is forcing function. We have written this as function of t^α purposely for ease. For example we will use in this paper $E_\alpha(ct^\alpha)$, $\sin_\alpha(ct^\alpha)$, t^α etc. are taken as forcing functions. There will be other functions in the derivations like $V(t^\alpha)$, $V(t^\alpha)$ all functions described with scaled variable that is t^α. Nevertheless the forcing functions can be written as simple $h(t)$ though.

In that paper [15] we found the following (theorems) which we will be using in this paper

(i) The fractional differential equation $\left({}_0^J D_t^\alpha - a\right)\left({}_0^J D_t^\alpha - b\right)y(t) = 0$ has solution of the form $y = AE_\alpha(at^\alpha) + BE_\alpha(bt^\alpha)$ where A and B are constants,

(ii) The fractional differential equation ${}_0^JD_t^{2\alpha}[y]-2a\left({}_0^JD^{\alpha}[y]\right)+a^2y=0$ has solution of the form $y=(At^{\alpha}+B)E_{\alpha}(at^{\alpha})$ where A and B are constants and

(iii) Solution of the fractional differential equation ${}_0^JD_t^{2\alpha}[y]-2a\left({}_0^JD_t^{\alpha}[y]\right)+(a^2+b^2)y=0$ is of the form $y=E_{\alpha}(at^{\alpha})[A\cos_{\alpha}(bt^{\alpha})+B\sin_{\alpha}(bt^{\alpha})]$ where A and B are constants.

From now we indicate Jumarie fractional derivative with start point of differentiation as 0 as ${}^JD^{\alpha}$ instead ${}_0^JD^{\alpha}$.

Theorem 1: If y_1 and y_2 are two solutions of the fractional differential equation $f\left({}^JD^{\alpha}\right)y=0$ then $c_1y_1+c_2y_2$ is also a solution, where c1 and c2 are arbitrary constants.

Proof: Since $f(D^{\alpha})y=0$ has solutions $y=y_1$ and $y=y_2$

$$f({}^JD^{\alpha})y_1=0 \qquad \text{and} \qquad f({}^JD^{\alpha})y_2=0$$

$$f({}^JD^{\alpha})(c_1y_1+c_2y_2)=c_1f({}^JD^{\alpha})y_1+c_2f({}^JD^{\alpha})y_2=0$$

Hence $c_1y_1+c_2y_2$ is also a solution of the given fractional differential equation.

Hence the theorem is proved.

Similarly, we can prove if $y_1, y_2, \ldots, y_n$ are solutions of the fractional differential equation $f({}^JD^{\alpha})y=0$ then $c_1y_1+c_2y_2+\ldots+c_ny_n$ is also a solution of it.

Theorem 2: If

$$f({}^JD^{\alpha})=({}^JD^{\alpha}-a_1)({}^JD^{\alpha}-a_2)({}^JD^{\alpha}-a_3)\ldots({}^JD^{\alpha}-a_n),$$
$$0<\alpha\le 1.$$

then solution of the homogeneous equation $f({}^JD^{\alpha})y=0$ is $y=\sum_{k=1}^{n}A_kE_{\alpha}(a_kt^{\alpha})$ where Ak's are arbitrary constants and all a_k are distinct.

Proof: Since Jumarie type fractional derivative of Mittag-Leffler function $y=E_{\alpha}(at^{\alpha})$ with a as a constant is ${}^JD^{\alpha}[y]=aE_{\alpha}(at^{\alpha})=ay, \quad 0<\alpha\le 1$. Thus solution of the differential equation ${}^JD^{\alpha}y=ay, \quad 0<\alpha\le 1$ is $y=AE_{\alpha}(at^{\alpha})$ where A is a constant [15].

Let $y=AE_{\alpha}(mt^{\alpha})\neq 0$ be a non-trivial trial solution of the differential equation $f({}^JD^{\alpha})y=0$ then ${}^JD^{\alpha}y=my$ or we write the following after subtracting a_iy from both the sides as demonstrated below

$${}^JD^{\alpha}y={}^JD^{\alpha}\left[AE_{\alpha}(mt^{\alpha})\right]=A\left(mE_{\alpha}(mt^{\alpha})\right)=my$$

$${}^JD^{\alpha}y-a_iy=my-a_iy$$

$$({}^JD^{\alpha}-a_i)y=(m-a_i)y$$

We apply the above result sequentially as demonstrated below

$$f\left({}^J D^\alpha\right)y = f({}^J D^\alpha)\left[AE_\alpha(mt^\alpha)\right]$$
$$=({}^J D^\alpha - a_1)({}^J D^\alpha - a_2)\ldots\left\{({}^J D^\alpha - a_n)\left[AE_\alpha(mt^\alpha)\right]\right\}$$
$$=({}^J D^\alpha - a_1)({}^J D^\alpha - a_2)\ldots({}^J D^\alpha - a_{n-1})$$
$$\left\{(m-a_n)\left[AE_\alpha(mt^\alpha)\right]\right\}$$
$$=({}^J D^\alpha - a_1)\left\{(m-a_2)(m-a_3)\ldots(m-a_n)\left[AE_\alpha(mt^\alpha)\right]\right\}$$
$$=(m-a_1)(m-a_2)\ldots(m-a_n)\left[AE_\alpha(mt^\alpha)\right]=\prod_{i=1}^{m}(m-a_i)y$$

Since $f({}^J D^\alpha)y=0$ we get

$$\prod_{i=1}^{n}(m-a_i)y=0 \tag{2.2}$$

Implying that $m=a_1,a_2,\ldots a_n$.

Hence the general solution is

$$y=A_1E_\alpha(a_1t^\alpha)+A_2E_\alpha(a_2t^\alpha)+\ldots+A_nE_\alpha(a_nt^\alpha)$$
$$=\sum_{k=1}^{n}A_kE(a_kt^\alpha)$$

Hence the theorem is proved.

The above theorem implies principal of superposition holds for the linear fractional differential equations (composed via Jumarie fractional derivative) also.

Note: In the above theorem if two or more roots of the equation (2.2) are equal or roots are complex then the solution [15] form is given below.

For $a_1=a_2$ and $a_3\neq a_4\neq\ldots\neq a_n$ then solution of the is

$$y=[A_1t^\alpha+A_2]E_\alpha(a_1t^\alpha)+A_3E_\alpha(a_3t^\alpha)$$
$$+\ldots+A_nE_\alpha(a_nt^\alpha)$$

For $a_1=a_2=a_3$ and $a_4\neq a_5\neq\ldots\neq a_n$ then the solution is

$$y = [A_1 t^{2\alpha} + A_2 t^{\alpha} + A_3] E_\alpha(a_1 t^\alpha) + A_4 E_\alpha(a_4 t^\alpha) + \ldots + A_n E_\alpha(a_n t^\alpha)$$

Where Ak's are arbitrary constants.

For $a_1, a_2 = a \pm ib$ and other are $a_3 \neq a_4 \neq \ldots \neq a_n$ then the solution is

$$y = [A_1 \cos_\alpha(bt^\alpha) + A_2 \sin_\alpha(bt^\alpha)] E_\alpha(at^\alpha) + A_3 E_\alpha(a_3 t^\alpha) + \ldots + A_n E_\alpha(a_n t^\alpha)$$

Thus solutions of linear homogeneous fractional differential equation with Jumarie fractional derivative is express in terms of Mittag-Leffler functions and fractional type sine and cosine series.

Now the question arises what will be solution of linear non-homogeneous fractional differential equations. The solution corresponding to the homogeneous equation will be called as the complementary function, it contains the arbitrary constants and this solution will be denoted by y_c. The other part, that is a solution which is free from integral constant, and depending on the forcing function will be called as Particular Integral (PI) and will be denoted by y_p. Thus the general solution will be $y = y_c + y_p$. We will develop simple method to evaluate Particular Integral.

α – ORDER NON-HOMOGENEOUS FRACTIONAL DIFFERENTIAL EQUATIONS

Consider the linear $\alpha-$ order non-homogeneous fractional differential equation with $0 < \alpha < 1$ for $y = 0$ for $t < 0$ of the following form,

$$\left({}^{J}D^{\alpha} - a\right) y = g(t^\alpha) \quad 0 < \alpha < 1 \tag{3.1}$$

The solution of the corresponding homogeneous part is [15]

$y_c = A_1 E_\alpha(mt^\alpha)$, A_1 is arbitary constant.

Multiply both side of equation (3.1) by $E_\alpha(-at^\alpha)$ as demonstrated below

$$\left[E_\alpha(-at^\alpha)\right]\left[({}^J D^\alpha - a)y\right] = g(t^\alpha)E_\alpha(-at^\alpha)$$

$$\left[E_\alpha(-at^\alpha)\right]\left[{}^J D^\alpha y - ay\right] = g(t^\alpha)E_\alpha(-at^\alpha)$$

$$\left[E_\alpha(-at^\alpha)\right]\left({}^J D^\alpha [y]\right) - ayE_\alpha(-at^\alpha)$$

$$= g(t^\alpha)E_\alpha(-at^\alpha)$$

$$\left[E_\alpha(-at^\alpha)\right]\left({}^J D^\alpha [y]\right) + y\left({}^J D^\alpha \left[E_\alpha(-at^\alpha)\right]\right)$$

$$= g(t^\alpha)E_\alpha(-at^\alpha)$$

$${}^J D^\alpha \left[yE_\alpha(-at^\alpha)\right] = g(t^\alpha)E_\alpha(-at^\alpha)$$

In the above steps we have used ${}_0^J D_t^\alpha \left[E_\alpha(-at^\alpha)\right] = -aE_\alpha(-at^\alpha)$. Now operating ${}^J D^{-\alpha}$ on both the sides of the obtained last expression in above derivation i.e.

$${}^J D^\alpha \left[yE_\alpha(-at^\alpha)\right] = g(t^\alpha)E_\alpha(-at^\alpha).$$

Also we add a constant A since Jumarie type derivative of a constant is zero and from here we get the following

$$\left[yE_\alpha(-at^\alpha)\right] = {}^J D^{-\alpha}\left[g(t^\alpha)E_\alpha(-at^\alpha)\right] + A$$

where A is a constant.

$$y = \left[E_\alpha(at^\alpha)\right]\left[{}^J D^{-\alpha}\left(g(t^\alpha)E_\alpha(-at^\alpha)\right) + A\right] \tag{3.2}$$

Or

$$y = AE_\alpha(at^\alpha) + \left[E_\alpha(at^\alpha)\right]\left[{}^J D^{-\alpha}\left(g(t^\alpha)E_\alpha(-at^\alpha)\right)\right]$$

the first part corresponds to solution of corresponding homogeneous equation, that is $y_c = AE_\alpha(at^\alpha)$ and the other part $y_p = \left[E_\alpha(at^\alpha)\right]\left[{}^J D^{-\alpha}\left(g(t^\alpha)E_\alpha(-at^\alpha)\right)\right]$ corresponds to the effect of non-homogeneous part and free from integral constant, but depending on the nature of forcing function, this part is named as Particular Integral (PI) as in case of classical differential equations. Now we take several forms of forcing function.

Particular Integral for $g(t^{\alpha}) = E_{\alpha}(ct^{\alpha})$

Here consider the linear first order non-homogeneous fractional differential equation of order α with $0 < \alpha \le 1$ with $y = 0$ for $t < 0$

$$({}^{J}D^{\alpha} - a)y = g(t^{\alpha}) \qquad g(t^{\alpha}) = E_{\alpha}(ct^{\alpha}), \qquad c \neq a.$$

then the particular integral (PI) described in the previous section is

$$y_p = \left[E_{\alpha}(at^{\alpha})\right]\left[{}^{J}D^{-\alpha}\left(g(t^{\alpha})E_{\alpha}(-at^{\alpha})\right)\right]$$

Putting $g(t^{\alpha}) = E_{\alpha}(ct^{\alpha})$ in above we get the following

$$\begin{aligned} y_p &= \left[E_{\alpha}(at^{\alpha})\right]\left[{}^{J}D^{-\alpha}\left(E_{\alpha}(ct^{\alpha})E_{\alpha}(-at^{\alpha})\right)\right] \\ &= \left[E_{\alpha}(at^{\alpha})\right]\left[{}^{J}D^{-\alpha}\left(E_{\alpha}((c-a)t^{\alpha})\right)\right] \\ &= \left[E_{\alpha}(at^{\alpha})\right]\left(\frac{1}{c-a}\right)\left[E_{\alpha}((c-a)t^{\alpha})\right] \\ &= \frac{1}{c-a}E_{\alpha}(ct^{\alpha}) \end{aligned}$$

For $c = a$, P.I. is

$$\begin{aligned} y_p &= \left[E_{\alpha}(at^{\alpha})\right]\left[{}^{J}D^{-\alpha}\left(E_{\alpha}(at^{\alpha})E_{\alpha}(-at^{\alpha})\right)\right] \\ &= \left[E_{\alpha}(at^{\alpha})\right]\left({}^{J}D^{-\alpha}[1]\right) \end{aligned}$$

use ${}^{J}D^{-\alpha}[1] = \dfrac{t^{\alpha}}{\Gamma(1+\alpha)}$

$$= \frac{t^{\alpha}}{\Gamma(1+\alpha)}E_{\alpha}(at^{\alpha})$$

Short procedure for calculating Particular Integral for $g(t^{\alpha}) = E_{\alpha}(ct^{\alpha})$.

This procedure is similar and in conjugation with classical integer order calculus. In classical order calculus $\alpha = 1$. Hence the forced function reduce to $g(t) = \exp(ct)$. Therefore the particular integral will be

$$y_p = [\exp(at)]\left[D^{-1}(\exp(ct).\exp(-at))\right]$$

$$= [\exp(at)]\left[\frac{1}{D}(\exp((c-a)t))\right]$$

$$= [\exp(at)]\left(\frac{1}{c-a}\right)[\exp((c-a)t)]$$

$$= \left(\frac{1}{c-a}\right)[\exp(ct)] \qquad \text{for } c \neq a.$$

For $c = a$

$$y_p = [\exp(at)]\left[\frac{1}{D}(\exp(at)\exp(-at))\right]$$

$$= [\exp(at)]\left[\frac{1}{D}(1)\right] = t\exp(at)$$

Here we observe that the derivative operator D is replaced by c in the first case, i.e. for $c \neq a$. In the second case the derivative operator D is replaced by $D + a$. We can replace the fractional Jumarie derivative operator ${}^J D^{\alpha}$ by c for the first case $c \neq a$ and by $\left({}^J D^{\alpha}\right) + a$ for second case $c = a$ The short procedure as follows for Particular Integral that is,

For $c \neq a$, $\quad y_p = \left(\frac{1}{{}^J D^{\alpha} - a}\right)\left[E_{\alpha}(ct^{\alpha})\right]$

replace ${}^J D^{\alpha}$ by c

$$= \frac{1}{c-a} E_{\alpha}(ct^{\alpha})$$

For $c = a$, $\quad y_p = \left(\frac{1}{{}^J D^{\alpha} - a}\right)\left[E_{\alpha}(at^{\alpha})\right]$

replace ${}^J D^{\alpha}$ by ${}^J D^{\alpha} + a$

$$= \left[E_{\alpha}(at^{\alpha})\right]\left(\frac{1}{{}^J D^{\alpha} + a - a}[1]\right)$$

$$= \left[E_{\alpha}(at^{\alpha})\right]\left(\frac{1}{{}^J D^{\alpha}}[1]\right)$$

$$= \left[E_{\alpha}(at^{\alpha})\right]\left({}^J D^{-\alpha}[1]\right)$$

$$= \frac{t^{\alpha}}{\Gamma(1+\alpha)} E_{\alpha}(at^{\alpha}).$$

Hence the general solution of equation (3.1) is $y = y_c + y_p$

$$y = \begin{cases} AE_\alpha(at^\alpha) + \dfrac{1}{c-a} E_\alpha(ct^\alpha) & \text{for } c \neq a \\ AE_\alpha(at^\alpha) + \dfrac{t^\alpha}{\Gamma(1+\alpha)} E_\alpha(at^\alpha) & \text{for } c = a. \end{cases}$$

Particular Integral for $g(t^\alpha) = t^\alpha$

Again when $g(t^\alpha) = t^\alpha$ then the differential equation (3.1) becomes

$$\left({}^J D^\alpha - a\right) y = t^\alpha \tag{3.3}$$

The solution of the homogeneous part [15] that is $({}^J D^\alpha - a)y = 0$ is $y_c = A_1 E_\alpha(at^\alpha)$.

Let $y = V(t^\alpha)E_\alpha(at^\alpha)$ the solution of the corresponding non-homogeneous equation where $V(t^\alpha)$ is an unknown function of t^α. Then using the definition by Jumarie [16] that is

$$(u(x)v(x))^{(\alpha)} = u^{(\alpha)}(x)v(x) + u(x)v^{(\alpha)}(x)$$

We get the following

$$\begin{aligned} {}^J D^\alpha[y] &= {}^J D^\alpha\left[V(t^\alpha)E_\alpha(at^\alpha)\right] \\ &= \left[V(t^\alpha)\right]\left({}^J D^\alpha\left[E_\alpha(at^\alpha)\right]\right) \\ &\quad + \left[E_\alpha(at^\alpha)\right]\left({}^J D^\alpha\left[V(t^\alpha)\right]\right) \\ &= \left[V(t^\alpha)\right]\left[aE_\alpha(at^\alpha)\right] + \left[E_\alpha(at^\alpha)\right]\left({}^J D^\alpha\left[V(t^\alpha)\right]\right) \end{aligned}$$

putting this in (3.3) we get

$$({}^J D^\alpha - a)y = t^\alpha$$
$${}^J D^\alpha[y] - ay = t^\alpha$$
$$\left[V(t^\alpha)\right]\left[aE_\alpha(at^\alpha)\right] + \left[E_\alpha(at^\alpha)\right]\left({}^J D^\alpha\left[V(t^\alpha)\right]\right)$$
$$-a\left[V(t^\alpha)\right]\left[E_\alpha(at^\alpha)\right] = t^\alpha$$
$$\left[E_\alpha(at^\alpha)\right]\left({}^J D^\alpha\left[V(t^\alpha)\right]\right) = t^\alpha$$

Therefore we get

$${}^J D^\alpha\left[V(t^\alpha)\right] = \frac{t^\alpha}{E_\alpha(at^\alpha)} = t^\alpha E_\alpha(-at^\alpha)$$

We now apply fractional integration by parts by Jumarie formula [16] as depicted below

$$\int_0^x u(y)\,{}^J D^\alpha[v(y)](dy)^\alpha$$
$$= u(y)v(y)\big|_0^x - \int_0^x v(y)\,{}^J D^\alpha[u(y)](dy)^\alpha$$

Here we mention that the symbol $\int f(x)(dx)^\alpha \equiv {}^J D^{-\alpha} f(x)$ implies Jumarie fractional integration as defined in section-2. We will use also [15] derived expression that is

$${}^J D^{-\alpha}\left[E_\alpha(-at^\alpha)\right] = (-a^{-1})\left[E_\alpha(-at^\alpha)\right],$$

in the following derivation.

$$V(t^{\alpha}) = A + {}^{J}D^{-\alpha}\left[t^{\alpha}E_{\alpha}(-at^{\alpha})\right]$$

$$= A + \int_{0}^{t} \tau^{\alpha}E_{\alpha}(-a\tau^{\alpha})(d\tau)^{\alpha}$$

$$= A + \int_{0}^{t} (\tau^{\alpha})\left({}^{J}D^{\alpha}\left[\frac{E_{\alpha}(-a\tau^{\alpha})}{-a}\right]\right)(d\tau)^{\alpha}$$

$$= A + \tau^{\alpha}\frac{E_{\alpha}(-a\tau^{\alpha})}{-a}\Bigg]_{0}^{t}$$

$$-\int_{0}^{t}\frac{E_{\alpha}(-a\tau^{\alpha})}{-a}\left({}^{J}D_{\tau}^{\alpha}\left[\tau^{\alpha}\right]\right)(d\tau)^{\alpha}$$

use $\quad {}^{J}D_{x}^{\alpha}\left[x^{\alpha}\right] = \Gamma(1+\alpha)$

$$= A + \tau^{\alpha}\frac{E_{\alpha}(-a\tau^{\alpha})}{-a}\Bigg]_{0}^{t} - \int_{0}^{t}\frac{E_{\alpha}(-a\tau^{\alpha})}{-a}\Gamma(1+\alpha)(d\tau)^{\alpha}$$

$$= A + \tau^{\alpha}\frac{E_{\alpha}(-a\tau^{\alpha})}{-a}\Bigg]_{0}^{t} + \frac{\Gamma(1+\alpha)}{a}\left[\frac{E_{\alpha}(-a\tau^{\alpha})}{-a}\right]_{0}^{t}$$

$$= A - t^{\alpha}\frac{E_{\alpha}(-at^{\alpha})}{a} - \frac{\Gamma(1+\alpha)}{a^{2}}\left[E_{\alpha}(-at^{\alpha}) - 1\right]$$

A = constant.

Hence the general solution is

$$y = \left[V(t^{\alpha})\right]\left[E_{\alpha}(at^{\alpha})\right]$$

$$= -\frac{1}{a}\left(t^{\alpha} + \frac{\Gamma(1+\alpha)}{a}\right) + \left(A + \frac{\Gamma(1+\alpha)}{a^{2}}\right)E_{\alpha}(at^{\alpha})$$

$$= A_{1}E_{\alpha}(at^{\alpha}) - \frac{1}{a}\left(t^{\alpha} + \frac{\Gamma(1+\alpha)}{a}\right)$$

where $\quad A_{1} = A + \dfrac{\Gamma(1+\alpha)}{a^{2}}$

$y_c = A_1 E_\alpha(at^\alpha)$ the first part in above expression is solution of homogeneous equation and the second part of the above that is $y_p = -\frac{1}{a}\left(t^\alpha + \frac{\Gamma(1+\alpha)}{a}\right)$ is particular integral.

Short procedure for Calculating Particular Integral for $g(t^\alpha) = t^\alpha$

This procedure is similar and in conjugation with classical integer order calculus. Here for $\alpha = 1$, and $g(t) = t$, and the corresponding particular integral is

$$y_p = \frac{1}{D-a}t = -\frac{1}{a}\left(1-\frac{D}{a}\right)^{-1} t$$

$$= -\frac{1}{a}\left\{1+\frac{D}{a}+\frac{D^2}{a^2}+\dots\right\}t = -\frac{1}{a}\left(t+\frac{1}{a}\right)$$

In the same way we can have a short procedure as follows for Particular Integral that is,

$$y_p = \frac{1}{{}^J D^\alpha - a} t^\alpha = -\frac{1}{a}\left(1-\frac{{}^J D^\alpha}{a}\right)^{-1} t^\alpha$$

$$= -\frac{1}{a}\left(1+\frac{{}^J D^\alpha}{a}+\frac{{}^J D^{2\alpha}}{a^2}+\dots\right)t^\alpha = -\frac{1}{a}\left(t^\alpha + \frac{\Gamma(1+\alpha)}{a}\right)$$

In the above derivation

$${}^J D^{2\alpha}\left[t^\alpha\right] = {}^J D^\alpha\left[D^\alpha(t^\alpha)\right] = {}^J D^\alpha\left[\Gamma(1+\alpha)\right] = D^\alpha\left[C\right] = 0$$

is used. Thus all the Jumarie derivatives ${}^J D^{k\alpha}\left[t^\alpha\right] = 0$ for $k > 1$, where k is Natural number. Therefore we have discussed the solutions of non-homogeneous $\alpha-$ order differential equations for different forcing functions $g(t^\alpha)$.

Evaluation of $y_p = \frac{1}{{}^J D^{2\alpha} - a^2} \sin_\alpha(ct^\alpha)$, $g(t^\alpha) = \sin_\alpha(ct^\alpha)$ where $c^2 + a^2 \neq 0$

$\frac{1}{{}^J D^{2\alpha} - a^2}$ can be factorized as $\frac{1}{{}^J D^{2\alpha} - a^2} = \frac{1}{({}^J D^\alpha - a)({}^J D^\alpha + a)} = \frac{1}{2a}\left[\frac{1}{{}^J D^\alpha - a} - \frac{1}{{}^J D^\alpha + a}\right]$,and we use this in following derivation.

As in section 3.1 here we replace ${}^J D^\alpha$ by ic and by- ic for the operations $\frac{1}{{}^J D^\alpha - a} E_\alpha(ict^\alpha)$ and $\frac{1}{{}^J D^\alpha - a} E_\alpha(-ict^\alpha)$ respectively, as is demonstrated below.

$$y_p = \frac{1}{{}^J D^{2\alpha} - a^2} \sin_\alpha(ct^\alpha)$$
$$= \frac{1}{2a}\left[\frac{1}{{}^J D^\alpha - a} - \frac{1}{{}^J D^\alpha + a}\right] \sin_\alpha(ct^\alpha)$$
$$= \frac{1}{2a}\left[\frac{1}{{}^J D^\alpha - a} \sin_\alpha(ct^\alpha) - \frac{1}{{}^J D^\alpha + a} \sin_\alpha(ct^\alpha)\right]$$
$$\sin_\alpha(ct^\alpha) \overset{\text{def}}{=} \frac{1}{2i}\left[E_\alpha(ict^\alpha) - E_\alpha(-ict^\alpha)\right]$$

$$\frac{1}{{}^J D^\alpha - a} \sin_\alpha(ct^\alpha) = \frac{1}{2i}\left[\begin{array}{l} \frac{1}{{}^J D^\alpha - a} E_\alpha(ict^\alpha) \\ - \frac{1}{{}^J D^\alpha - a} E_\alpha(-ict^\alpha) \end{array}\right]$$
$$= \frac{1}{2i}\left[\frac{1}{ic - a} E_\alpha(ict^\alpha) - \frac{1}{-ic - a} E_\alpha(-ict^\alpha)\right]$$
$$\frac{1}{{}^J D^\alpha + a} \sin_\alpha(ct^\alpha) = \frac{1}{2i}\left[\begin{array}{l} \frac{1}{{}^J D^\alpha + a} E_\alpha(ict^\alpha) \\ - \frac{1}{{}^J D^\alpha + a} E_\alpha(-ict^\alpha) \end{array}\right]$$
$$= \frac{1}{2i}\left[\frac{1}{ic + a} E_\alpha(ict^\alpha) - \frac{1}{-ic + a} E_\alpha(-ict^\alpha)\right]$$

Therefore

$$\frac{1}{{}^{J}D^{2\alpha}-a^{2}}\sin_{\alpha}(ct^{\alpha})$$
$$=\frac{1}{4ai}\left[\begin{array}{l}\frac{1}{ic-a}E_{\alpha}(it^{\alpha})-\frac{1}{ic+a}E_{\alpha}(ict^{\alpha})\\ -\frac{1}{-ic-a}E_{\alpha}(-ict^{\alpha})+\frac{1}{-ic+a}E_{\alpha}(-ict^{\alpha})\end{array}\right]$$
$$=\frac{1}{4ai}\left[\begin{array}{l}\left(\frac{1}{ic-a}-\frac{1}{ic+a}\right)E_{\alpha}(ict^{\alpha})\\ -\left(\frac{1}{-ic-a}-\frac{1}{-ic+a}\right)E_{\alpha}(-ict^{\alpha})\end{array}\right]$$
$$=\frac{1}{2i}\left[\frac{1}{-c^{2}-a^{2}}E_{\alpha}(ict^{\alpha})-\frac{1}{-c^{2}-a^{2}}E_{\alpha}(-ict^{\alpha})\right]$$
$$=\left(\frac{1}{-c^{2}-a^{2}}\right)\left(\frac{1}{2i}\left[E_{\alpha}(ict^{\alpha})-E_{\alpha}(-ict^{\alpha})\right]\right)$$
$$=\frac{\sin_{\alpha}(ct^{\alpha})}{-c^{2}-a^{2}}$$

Similarly we get by following above procedure

$$\frac{1}{{}^{J}D^{2\alpha}-a^{2}}\cos_{\alpha}(ct^{\alpha})=\frac{\cos_{\alpha}(ct^{\alpha})}{-c^{2}-a^{2}}$$

Thus to find the particular integral $\frac{1}{{}^{J}D^{2\alpha}-a^{2}}\sin_{\alpha}(ct^{\alpha})$ replace ${}^{J}D^{2\alpha}$ by $-c^{2}$.

This procedure is similar and in conjugation with classical integer order calculus. In classical order calculus $\alpha=1$ hence the forced function reduce to $g(t)=\sin(ct)$. Therefore the particular integral will be

$$y_{p}=\frac{1}{{}^{J}D^{2}-a^{2}}\sin(ct)$$
$$=\frac{1}{2a}\left[\frac{1}{{}^{J}D-a}-\frac{1}{{}^{J}D+a}\right]\sin(ct)$$
$$=\frac{1}{2a}\left[\frac{1}{{}^{J}D-a}\sin(ct)-\frac{1}{{}^{J}D+a}\sin(ct)\right]$$

$$\sin(ct)=\frac{1}{2i}\left[\exp(ict)-\exp(-ict)\right]$$

$$\frac{1}{{}^{J}D-a}\sin(ct)=\frac{1}{2i}\begin{bmatrix}\dfrac{1}{{}^{J}D-a}\exp(ict)\\ -\dfrac{1}{{}^{J}D-a}\exp(-ict)\end{bmatrix}$$

$$=\frac{1}{2i}\left[\frac{1}{ic-a}\exp(ict)-\frac{1}{-ic-a}\exp(-ict)\right]$$

$$\frac{1}{{}^{J}D+a}\sin(ct)=\frac{1}{2i}\begin{bmatrix}\dfrac{1}{{}^{J}D+a}\exp(ict)\\ -\dfrac{1}{{}^{J}D+a}\exp(-ict)\end{bmatrix}$$

$$=\frac{1}{2i}\left[\frac{1}{ic+a}\exp(ict)-\frac{1}{-ic+a}\exp(-ict)\right]$$

Therefore

$$\frac{1}{{}^{J}D^{2}-a^{2}}\sin(ct)$$

$$=\frac{1}{4ai}\begin{bmatrix}\dfrac{1}{ic-a}\exp(ict)-\dfrac{1}{ic+a}\exp(ict)\\ -\dfrac{1}{-ic-a}\exp(-ict)+\dfrac{1}{-ic+a}\exp(-ict)\end{bmatrix}$$

$$=\frac{1}{4ai}\begin{bmatrix}\left(\dfrac{1}{ic-a}-\dfrac{1}{ic+a}\right)\exp(ict)\\ -\left(\dfrac{1}{-ic-a}-\dfrac{1}{-ic+a}\right)\exp(ict)\end{bmatrix}$$

$$=\frac{1}{2i}\left[\frac{1}{-c^{2}-a^{2}}\exp(ict)-\frac{1}{-c^{2}-a^{2}}\exp(-ict)\right]$$

$$=\left(\frac{1}{-c^{2}-a^{2}}\right)\left(\frac{1}{2i}\left[\exp(ict)-\exp(-ict)\right]\right)$$

$$=\frac{\sin(ct)}{-c^{2}-a^{2}}$$

$2\alpha-$ ORDER NON-HOMOGENEOUS FRACTIONAL DIFFERENTIAL EQUATIONS

General formulation of non-homogeneous fractional differential equation of

2α – order is $\quad {}^{J}D^{2\alpha}[y] + p\left({}^{J}D^{\alpha}[y]\right) + qy = g(t^{\alpha})$

$0 < \alpha < 1 \quad y(t) = 0 \quad \text{for} \quad t < 0$

Where p and q are constant here. Consider the $2\alpha-$ order non-homogeneous fractional differential equation

$$f({}^{J}D^{\alpha})y = g(t^{\alpha})$$

Where $f({}^{J}D^{\alpha}) = ({}^{J}D^{\alpha} - a)({}^{J}D^{\alpha} - b)$ then solution of the non-homogeneous part that is $f({}^{J}D^{\alpha}) = 0$ given by $y_c = AE_{\alpha}(at^{\alpha}) + BE_{\alpha}(bt^{\alpha})$ [15].

Use of Method of Un-determinant Coefficient Method to Calculate the Particular Integrals for Different Functional Forms of $g(t^{\alpha})$

For $g(t^{\alpha}) = E_{\alpha}(ct^{\alpha})$ we have the given equation is

$$\begin{aligned} f({}^{J}D^{\alpha})y &= ({}^{J}D^{\alpha} - a)({}^{J}D^{\alpha} - b)y \\ &= \left[{}^{J}D^{2\alpha} - (a+b)\left({}^{J}D^{\alpha}\right) + ab\right](y) \\ &= E_{\alpha}(ct^{\alpha}) \qquad \text{for } c \neq a \neq b \end{aligned} \tag{4.1}$$

Here let the particular integral be $y_p = PE_{\alpha}(ct^{\alpha})$ where P is constant.

Then

$${}^{J}D^{\alpha}\left[y_p\right] = PcE_{\alpha}(ct^{\alpha}) \quad {}^{J}D^{2\alpha}\left[y_p\right] = Pc^2E_{\alpha}(ct^{\alpha})$$

And putting in the given equation (4.1) we get the following

$$\left[{}^{J}D^{2\alpha}-(a+b)\left({}^{J}D^{\alpha}\right)+ab\right](y_p)=E_\alpha(ct^\alpha)$$
$${}^{J}D^{2\alpha}\left[y_p\right]-(a+b)\left({}^{J}D^{\alpha}\left[y_p\right]\right)+aby_p=E_\alpha(ct^\alpha)$$
$$Pc^2E_\alpha(ct^\alpha)-(a+b)PcE_\alpha(ct^\alpha)$$
$$+abPE_\alpha(ct^\alpha)=E_\alpha(ct^\alpha)$$
$$Pc^2-(a+b)Pc+abP=1$$
$$P\left[c^2-(a+b)c+ab\right]=1$$
$$P\left[(c-a)(c-b)\right]=1$$

Therefore

$$P=\frac{1}{(c-a)(c-b)}$$

and consequently the Particular integral is

$$y_p=\frac{1}{(c-a)(c-b)}E_\alpha(ct^\alpha)$$

Hence the general solution is

$$y=AE_\alpha(at^\alpha)+BE_\alpha(bt^\alpha)+\frac{1}{(c-a)(c-b)}E_\alpha(ct^\alpha) \tag{4.2}$$

For $c=a\neq b$ implying $(c-a)(c-b)=0$ then the solution (4.2) does not exists. In this case the fractional differential equation is

$$f({}^{J}D^{\alpha})y=({}^{J}D^{\alpha}-a)({}^{J}D^{\alpha}-b)y$$
$$=\left[{}^{J}D^{2\alpha}-(a+b)\left({}^{J}D^{\alpha}\right)+ab\right]y=E_\alpha(at^\alpha) \tag{4.3}$$

If we consider $y_p=PE_\alpha(at^\alpha)$ in this case also then putting in (4.3) we get $0=E_\alpha(at^\alpha)$ which is free from P i.e P is non-determinable. This form of PI is not suitable here; consider the modified form as following

$$y_p=Pt^\alpha E_\alpha(at^\alpha)$$

Then

$$^{J}D^{\alpha}\left[y_p\right]=P\left[at^{\alpha}E_{\alpha}(at^{\alpha})+\Gamma(1+\alpha)E_{\alpha}(at^{\alpha})\right]$$

$$^{J}D^{2\alpha}\left[y_p\right]=P\left[a^2t^{\alpha}E_{\alpha}(at^{\alpha})+2a\Gamma(1+\alpha)E_{\alpha}(at^{\alpha})\right]$$

And putting in (4.3) we get the following

$$Pa^2t^{\alpha}+2aP\Gamma(1+\alpha)-(a+b)Pat^{\alpha}$$

$$-(a+b)P\Gamma(1+\alpha)+Pabt^{\alpha}=1$$

$$P=\frac{1}{(a-b)\Gamma(1+\alpha)}$$

$$y_p=\frac{t^{\alpha}}{(a-b)\Gamma(1+\alpha)}E_{\alpha}(at^{\alpha})$$

In this case the general solution of the fractional differential equation is of following form

$$y=AE_{\alpha}(at^{\alpha})+BE_{\alpha}(bt^{\alpha})+\frac{t^{\alpha}}{(a-b)\Gamma(1+\alpha)}E_{\alpha}(at^{\alpha})$$

When $c=a=b$ then

$$f(^{J}D^{\alpha})y=(^{J}D^{\alpha}-a)^2y=E_{\alpha}(at^{\alpha}) \tag{4.4}$$

and take the particular integral in the form

$$y_p=Pt^{2\alpha}E_{\alpha}(at^{\alpha})$$

then

$$D^{\alpha}y_p=Pat^{2\alpha}E_{\alpha}(at^{\alpha})+Pt^{\alpha}\frac{\Gamma(1+2\alpha)}{\Gamma(1+\alpha)}E_{\alpha}(at^{\alpha})$$

$$^{J}D^{2\alpha}\left[y_p\right]=Pa^2t^{2\alpha}E_{\alpha}(at^{\alpha})$$

$$+2aPt^{\alpha}\frac{\Gamma(1+2\alpha)}{\Gamma(1+\alpha)}E_{\alpha}(at^{\alpha})+P\Gamma(1+2\alpha)E_{\alpha}(at^{\alpha})$$

Putting this in (4.4) and after simplification we get

$$P = \frac{1}{\Gamma(1+2\alpha)}$$

$$y_p = \frac{t^{2\alpha}}{\Gamma(1+2\alpha)} E_\alpha(at^\alpha)$$

In this case the general solution will be of following form

$$y = (At^\alpha + B)E_\alpha(at^\alpha) + \frac{t^{2\alpha}}{\Gamma(1+2\alpha)} E_\alpha(at^\alpha)$$

Thus we can summarize the result as a theorem in the following form

Theorem: The differential equation the $f({}^J D^\alpha)y = E_\alpha(ct^\alpha)$ has particular integral

$$\frac{1}{f(c)} E_\alpha(ct^\alpha) \qquad \text{for} \qquad f(c) \neq 0$$

(i) When $f({}^J D^\alpha) = ({}^J D^\alpha - a)({}^J D^\alpha - b)$ for $c \neq a \neq b$ then solution of the fractional differential equation will be

$$y = AE_\alpha(at^\alpha) + BE_\alpha(bt^\alpha) + \frac{1}{(c-a)(c-b)} E_\alpha(ct^\alpha)$$

(ii) When $f({}^J D^\alpha) = ({}^J D^\alpha - a)({}^J D^\alpha - b)$ for $c = a \neq b$ then solution of the fractional differential equation will be

$$y = AE_\alpha(at^\alpha) + BE_\alpha(bt^\alpha) + \frac{t^\alpha}{(a-b)\Gamma(1+\alpha)} E_\alpha(at^\alpha)$$

(iii) When $f({}^J D^\alpha) = ({}^J D^\alpha - a)({}^J D^\alpha - b)$ for $c = a = b$ then solution of the fractional differential equation will be

$$y = (At^\alpha + B)E_\alpha(at^\alpha) + \frac{t^{2\alpha}}{\Gamma(1+2\alpha)} E_\alpha(at^\alpha)$$

Use of Direct Method to Calculate the Particular Integrals for Different Functional Format of $g(t^{\alpha})$

Using the direct method as describe in section 3.1 we can easily calculate the Particular integrals for different functional format of $g(t^{\alpha})$.

For $g(t^{\alpha})=E_{\alpha}(ct^{\alpha})$ we have

$$\begin{aligned}
y_p &= \frac{1}{f({}^J D^{\alpha})} E_{\alpha}(ct^{\alpha}) \\
&= \frac{1}{({}^J D^{\alpha}-a)({}^J D^{\alpha}-b)} E_{\alpha}(ct^{\alpha}) \\
&= \frac{1}{a-b}\left[\frac{1}{({}^J D^{\alpha}-a)} - \frac{1}{({}^J D^{\alpha}-b)}\right] E_{\alpha}(ct^{\alpha}) \\
&= \frac{1}{a-b}\left[\frac{1}{(c-a)} - \frac{1}{(c-b)}\right] E_{\alpha}(ct^{\alpha}) \qquad \text{for } c \neq a \neq b \\
&= \frac{1}{(c-b)(c-a)} E_{\alpha}(ct^{\alpha}) \qquad \text{for } c \neq a \neq b.
\end{aligned}$$

In this case the general solution is

$$y = AE_{\alpha}(at^{\alpha}) + BE_{\alpha}(bt^{\alpha}) + \frac{1}{(c-b)(c-a)} E_{\alpha}(ct^{\alpha})$$

$\text{for } c \neq a$

For $c = a$

$$\begin{aligned}
y_p &= \frac{1}{f({}^J D^{\alpha})} E_{\alpha}(at^{\alpha}) \\
&= \frac{1}{({}^J D^{\alpha}-a)({}^J D^{\alpha}-b)} E_{\alpha}(at^{\alpha}) \\
&= \frac{1}{a-b}\left[\frac{1}{({}^J D^{\alpha}-a)} - \frac{1}{({}^J D^{\alpha}-b)}\right] E_{\alpha}(at^{\alpha}) \\
&= \frac{1}{a-b}\left[\frac{t^{\alpha} E_{\alpha}(at^{\alpha})}{\Gamma(1+\alpha)} - \frac{1}{(a-b)}\right] E_{\alpha}(at^{\alpha})
\end{aligned}$$

In this case the second part will be adjusted in the complementary function and hence the general solution is

$$y = AE_\alpha(at^\alpha) + BE_\alpha(bt^\alpha) + \frac{t^\alpha}{(a-b)\Gamma(1+\alpha)}E_\alpha(at^\alpha)$$

For $c = a = b$

$$\begin{aligned} y_p &= \frac{1}{f({}^J D^\alpha)}E_\alpha(at^\alpha) \\ &= \frac{1}{\left({}^J D^\alpha - a\right)^2}E_\alpha(at^\alpha) \\ &= \left[\frac{1}{\left({}^J D^\alpha - a\right)^2}\right]E_\alpha(at^\alpha) \\ &= E_\alpha(at^\alpha)\frac{1}{\left({}^J D^\alpha + a - a\right)^2}[1] \\ &= \frac{t^{2\alpha}E_\alpha(at^\alpha)}{\Gamma(1+2\alpha)}. \end{aligned}$$

In this case the general solution will be

$$y = (At^\alpha + B)E_\alpha(at^\alpha) + \frac{t^{2\alpha}}{\Gamma(1+2\alpha)}E_\alpha(at^\alpha)$$

In generalized case for any polynomial type function $f({}^J D^\alpha)$ the particular integral is

$$y_p = \frac{1}{f(D^\alpha)}E_\alpha(ct^\alpha) = \frac{1}{f(c)}E_\alpha(ct^\alpha) \quad \text{for} \quad f(c) \neq 0.$$

For $f(c) = 0$ $f({}^J D^\alpha)$ must contain a factor of the form $\left({}^J D^\alpha - c\right)^k$, k-positive integer i.e.

$$f({}^J D^\alpha) = \left({}^J D^\alpha - c\right)^k \phi({}^J D^\alpha) \text{ with } \phi(c) \neq 0$$

Then

$$y_p = \frac{1}{({}^J D^{\alpha} - c)^k \phi({}^J D^{\alpha})} E_{\alpha}(ct^{\alpha})$$
$$= \frac{E_{\alpha}(ct^{\alpha})}{\phi(c)} \frac{1}{\left({}^J D^{\alpha} - c\right)^k}[1]$$
$$= \frac{E_{\alpha}(ct^{\alpha})}{\phi(c)} \frac{t^{k\alpha}}{\Gamma(1+k\alpha)}$$

Again if $f({}^J D^{\alpha}) = v(t^{\alpha})E_{\alpha}(ct^{\alpha})$ Here using Leibnitz rule of fractional derivative on (Jumarie type fractional derivative) we get the following steps

$${}^J D^{\alpha}\left(v(t^{\alpha})E_{\alpha}(ct^{\alpha})\right)$$
$$= \left({}^J D^{\alpha}\left[v(t^{\alpha})\right]\right)\left[E_{\alpha}(ct^{\alpha})\right] + cv(t^{\alpha})E_{\alpha}(ct^{\alpha})$$
$$= ({}^J D^{\alpha} + c)v(t^{\alpha})E_{\alpha}(ct^{\alpha})$$

operate ${}^J D^{-\alpha}$ on both sides

$$v(t^{\alpha})E_{\alpha}(ct^{\alpha}) = {}^J D^{-\alpha}\left[({}^J D^{\alpha} + c)v(t^{\alpha})E_{\alpha}(ct^{\alpha})\right]$$

Let $\quad V(t^{\alpha}) = ({}^J D^{\alpha} + c)v(t^{\alpha})$

then $\quad v(t^{\alpha}) = \frac{1}{{}^J D^{\alpha} + c} V(t^{\alpha})$

$${}^J D^{-\alpha}\left[V(t^{\alpha})E_{\alpha}(ct^{\alpha})\right] = \frac{E_{\alpha}(ct^{\alpha})}{{}^J D^{\alpha} + c} V(t^{\alpha})$$

Thus we obtain

$${}^J D^{-\alpha}\left[V(t^{\alpha})E_{\alpha}(ct^{\alpha})\right] = \frac{E_{\alpha}(ct^{\alpha})}{D^{\alpha} + c} V(t^{\alpha})$$

$$\frac{1}{{}^J D^{\alpha}} V(t^{\alpha})E_{\alpha}(ct^{\alpha}) = \frac{E_{\alpha}(ct^{\alpha})}{{}^J D^{\alpha} + c} V(t^{\alpha})$$

Thus in generalized case we have

$$\frac{1}{f({}^J D^{\alpha})} V(t^{\alpha})E_{\alpha}(ct^{\alpha}) = \frac{E_{\alpha}(ct^{\alpha})}{f({}^J D^{\alpha} + c)} V(t^{\alpha})$$

SOLUTION THE FRACTIONAL DIFFERENTIAL-APPLICATION OF METHOD DERIVED

Example 1: We take the following fractional differential equation $y(t)=0$ for $t<0$

$$\left({}^{J}D^{2/3}[y]\right)-5\left(D^{1/3}[y]\right)+6y=t^{2}$$

Solution: Solution of the corresponding homogeneous equation is [15]

$$y_c = AE_{1/3}(2t^{1/3}) + BE_{1/3}(3t^{1/3})$$

The particular integral calculation is done in following steps

$$y_p = \frac{1}{({}^{J}D^{1/3}-2)({}^{J}D^{1/3}-3)}t^{6/3}$$

$$=\left[\frac{1}{({}^{J}D^{1/3}-3)}-\frac{1}{({}^{J}D^{1/3}-2)}\right]t^{6/3}$$

$$=\left[-\frac{1}{3}\left(1-\frac{{}^{J}D^{1/3}}{3}\right)^{-1}+\frac{1}{2}\left(1-\frac{{}^{J}D^{1/3}}{2}\right)^{-1}\right]t^{6/3}$$

$$=-\frac{1}{3}\left(1-\frac{{}^{J}D^{1/3}}{3}+\frac{{}^{J}D^{2/3}}{9}-\frac{{}^{J}D^{3/3}}{27}+....\right)t^{6/3}$$

$$+\frac{1}{2}\left(1-\frac{{}^{J}D^{1/3}}{2}+\frac{{}^{J}D^{2/3}}{4}-\frac{{}^{J}D^{3/3}}{8}+....\right)t^{6/3}$$

$$=\frac{1}{6}t^{6/3}+\frac{3^2-2^2}{3^2.2^2}\frac{\Gamma(1+2)}{\Gamma\left(1+\frac{5}{3}\right)}t^{5/3}-\frac{3^3-2^3}{3^3.2^3}\frac{\Gamma(1+2)}{\Gamma\left(1+\frac{4}{3}\right)}t^{4/3}$$

$$+\frac{3^4-2^4}{3^4.2^4}\frac{\Gamma(1+2)}{\Gamma\left(1+\frac{3}{3}\right)}t^{3/3}+\frac{3^5-2^5}{3^5.2^5}\frac{\Gamma(1+2)}{\Gamma\left(1+\frac{2}{3}\right)}t^{2/3}$$

$$+\frac{3^6-2^6}{3^6.2^6}\frac{\Gamma(1+2)}{\Gamma\left(1+\frac{1}{3}\right)}t^{1/3}+\frac{3^7-2^7}{3^7.2^7}\frac{\Gamma(1+2)}{\Gamma(1)}$$

Hence the general solution is

$$y = AE_{\frac{1}{3}}(2t^{\frac{1}{3}}) + BE_{\frac{1}{3}}(3t^{\frac{1}{3}}) + \frac{1}{6}t^{\frac{6}{3}}$$
$$+\frac{10}{6^2\Gamma\left(\frac{8}{3}\right)}t^{\frac{5}{3}} - \frac{38}{6^3\Gamma\left(\frac{7}{3}\right)}t^{\frac{4}{3}} + \frac{130}{6^4\Gamma\left(\frac{6}{3}\right)}t^{\frac{3}{3}}$$
$$+\frac{422}{6^5\Gamma\left(\frac{5}{3}\right)}t^{\frac{2}{3}} + \frac{1330}{6^6\Gamma\left(\frac{4}{3}\right)}t^{\frac{1}{3}} + \frac{4118}{6^7}$$

Where A and B are arbitrary constants.

Example 2: Consider the fractional order forced differential equation $y(t) = 0$ for $t < 0$

$$\left({}^{J}D^{2\alpha}[y]\right) + \omega^2 y = F\cos_\alpha(at^\alpha) \quad a \neq \omega$$

Solution: Here solution of the corresponding homogeneous equation $\left({}^{J}D^{2\alpha}[y]\right) + \omega^2 y = 0$ is [15]

$$y_c = A\cos_\alpha(\omega t^\alpha) + B\sin_\alpha(\omega t^\alpha)$$

The particular integral is

$$y_p = \frac{1}{D^{2\alpha} + \omega^2}F\cos_\alpha(at^\alpha) = \frac{1}{-a^2 + \omega^2}F\cos_\alpha(at^\alpha)$$

Hence the general solution is

$$y = A\cos_\alpha(\omega t^\alpha) + B\sin_\alpha(\omega t^\alpha) + \frac{1}{-a^2 + \omega^2}F\cos_\alpha(at^\alpha)$$

For particular integral we replaced ${}^{J}D^\alpha \equiv ia$ so ${}^{J}D^{2\alpha} \equiv (ia)^2 = -a^2$

Example 3: Take the fractional order damped-forced differential equation $y(t) = 0$ for $t < 0$

$$\left({}^{J}D^{2\alpha}[y]\right)+2c\left(D^{\alpha}[y]\right)+(c^2+\omega^2)y$$
$$=F\cos_{\alpha}(at^{\alpha})\quad a\neq\omega$$

Solution: Here solution of the corrεsponding homogeneous equation [15]

$$\left({}^{J}D^{2\alpha}[y]\right)+2c\left(D^{\alpha}[y]\right)+(c^2-\omega^2)y=0$$

Is

$$y_c=\left[E_{\alpha}(ct^{\alpha})\right]\left[A\cos_{\alpha}(\omega t^{\alpha})+B\sin_{\alpha}(\omega t^{\alpha})\right]$$

The particular integral is

$$y_p=\frac{1}{{}^{J}D^{2\alpha}+2c\left({}^{J}D^{\alpha}\right)+(c^2+\omega^2)}F\cos_{\alpha}(at^{\alpha})$$
$$=\frac{1}{2c\left({}^{J}D^{\alpha}\right)+(c^2-a^2+\omega^2)}F\cos_{\alpha}(at^{\alpha}),$$

here replace ${}^{J}D^{2\alpha}$ by $-a^2$
multiply the number pr and denominator by

$$(c^2-a^2+\omega^2)-2c\left({}^{J}D^{\alpha}\right)$$

$$y_p=\frac{(c^2-a^2+\omega^2)-2c\left({}^{J}D^{\alpha}\right)}{(c^2-a^2+\omega^2)^2-4c^2\left({}^{J}D^{2\alpha}\right)}F\cos_{\alpha}(at^{\alpha}),$$

here again replace ${}^{J}D^{2\alpha}$ by $-a^2$

$$=F\frac{(c^2-a^2+\omega^2)\cos_{\alpha}(at^{\alpha})-2ca\sin_{\alpha}(at^{\alpha})}{(c^2-a^2+\omega^2)^2+4c^2a^2}$$

Hence the general solution is

$$y = \left(A\cos_\alpha(\omega t^\alpha) + B\sin_\alpha(\omega t^\alpha)\right)\left[E_\alpha(ct^\alpha)\right]$$
$$+F\frac{(c^2 - a^2 + \omega^2)\cos_\alpha(at^\alpha) - 2ca\sin_\alpha(at^\alpha)}{(c^2 - a^2 + \omega^2)^2 + 4c^2a^2}$$

CONCLUSIONS

In this paper we have developed a method to solve the linear fractional non-homogeneous fractional differential equations, composed by Jumarie type derivative. The solutions are obtained here in terms of Mittag-Leffler function and fractional sine fractional cosine functions. Here we have proved via usage of Jumarie fractional derivative operator that for obtaining the particular integrals for several forcing functions scaled in function of variable t^α eases the method, and we obtain conjugation with classical method to solve classical non-homogeneous differential equations. The short cut rules, that are developed here in this paper to replace the operator D^α or operator $D^{2\alpha}$ as were used in classical calculus, gives ease and advantage in evaluating particular integrals. These techniques obtained herein this paper is remarkable to study fractional dynamic systems, and eases to get solution in terms of Mittag-Leffler, and fractional-trigonometric functions as in conjugation with exponential and normal trigonometric function for normal integer order calculus. Therefore this developed method is useful and advantageous as it is having conjugation with the classical methods of solving non-homogeneous fractional linear differential equations composed via Jumarie fractional derivative, and is also useful in understanding physical systems described by FDE.

ACKNOWLEDGEMENT

Acknowledgments are to Board of Research in Nuclear Science (BRNS), Department of Atomic Energy Government of India for financial assistance received through BRNS research project no. 37(3)/14/46/2014-BRNS with BSC BRNS, title “Characterization of unreachable (Holderian) functions via Local Fractional Derivative and Deviation Function.

REFERENCES

1. S. Das. “Functional Fractional Calculus”, 2nd Edition, Springer-Verlag Germany. (2011).
2. E. Ahmed, A.M.A. El-Sayed, H.A.A. El-Saka. “Equilibrium points, stability and numerical solutions of fractional-order predator-prey and rabies models”. J Math Anal Appl 2007; 325:54-53. (2007).

3. A. Alsaedi, S. K Ntouyas, R. P Agarwal, B. Ahmad. "On Caputo type sequential fractional differential equations with nonlocal integral boundary conditions". Advances in Difference Equations 2015; 33. 1-12.

4. K. S. Miller, B. Ross. "An Introduction to the Fractional Calculus and Fractional Differential Equations". John Wiley & Sons, New York, NY, USA; (1993).

5. S. S. Ray, R.K. Bera. "An approximate solution of a nonlinear fractional differential equation by Adomian decomposition method". Applied Mathematics and Computation. 2005; 167. 561-571. (2005).

6. I. Podlubny "Fractional Differential Equations", Mathematics in Science and Engineering, Academic Press, San Diego, Calif, USA. 1999; 198. (1999).

7. K. Diethelm. "The analysis of Fractional Differential equations". Springer-Verlag, Germany. (2010).

8. A. Kilbas, H. M. Srivastava, J.J. Trujillo, "Theory and Applications of Fractional Differential Equations". North-Holland Mathematics Studies, Elsevier Science, Amsterdam, the Netherlands. 2006; 204. (2006).

9. B. Ahmad, J.J. Nieto. "Riemann-Liouville fractional integro-differential equations with fractional nonlocal integral boundary conditions". Bound. Value Probl. 2011; 36. (2011).

10. B. Zheng. "Exp-Function Method for Solving Fractional Partial Differential Equations". Hindawi Publishing Corporation. The Scientific-World Journal. 2013; 1-8.

11. O. Abdulaziz, I. Hashim, S. Momani, "Application of homotopy-perturbation method to fractional IVPs", J. Comput. Appl. Math. 2008; 216. 574-584. (2008).

12. G.C. Wu, E.W.M. Lee, "Fractional Variational Iteration Method and its Application", Phys. Lett. A. 2010; 374 2506-2509. (2010).

13. D. Nazari, S. Shahmorad. "Application of the fractional differential transform method to fractional-order integro-differential equations with nonlocal boundary conditions". Journal of Computational and Applied Mathematics. 2010; 234(3). 883-891. (2010).

14. S. Zhang and H. Q. Zhang, "Fractional sub-equation method and its applications to nonlinear fractional PDEs", Phys. Lett. A, 2011; 375. 1069-1073. (2011).

15. U. Ghosh, S. Sengupta, S. Sarkar and S. Das. "Analytic solution of linear fractional differential equation with Jumarie derivative in term of Mittag-Leffler function". American Journal of Mathematical Analysis. 2015; 3(2). 32-38. (2015).

16. G. Jumarie. "Modified Riemann-Liouville derivative and fractional Taylor series of non-differentiable functions-Further results", Computers and Mathematics with Applications, 2006; 51. 1367-1376. (2006).

17. U. Ghosh, S. Sengupta, S. Sarkar and S. Das. "Characterization of non-differentiable points of a function by Fractional derivative of Jumarie type". European Journal of Academic Essays 2015; 2(2): 70-86. (2015).

18. G. M. Mittag-Leffler. "Sur la nouvelle function Eα (x)", C. R. Acad. Sci. Paris, (Ser. II). 1903; 137, 554-558. (1903).

3

LINEAR MODEL ANALYSIS OF OBSERVATIONAL DATA IN THE SENSE OF LEAST –SQUARES CRITERION

J.A. Basabrain*

* Department of Statistics College of Science for Girls, King Abdulaziz University, Jeddah, Saudi Arabia

ABSTRACT

The present paper is devoted for the following goals: To develop an algorithm for model analysis of observational data in the sense of the least –squares criterion with full error analysis.. By this algorithm one computes, all the solutions with their variances, the variance of the fit, the average square distance between the least square solution and the exact solution, and graphical representation between the row and the fitted data. Mathematica module of the algorithm was established, through five points, its purpose –input - output –needed procedures and the list of the module. By this paper we have been tried to produce an error controlled algorithm of the least squares method for observational data.

INTRODUCTION

One of the hallmarks of modern astrophysics is the extreme rise in the amount of data available and the prevalence of images and other data in electronic form particularly that available to the public via the Internet. The present situation is in marked contrast to the view given by Kolb and Turner [1] more than 25 years ago who lamented that in cosmology at that time there was a paucity of data. There are now so many people involved with data reduction and analysis that in the last decadal survey there was a call for the formation of a new area of astronomy and astrophysics called astroinformatics [3]. Latest information on this rapidly emerging discipline can be found with listings on the Internet through a Google or similar search under that name.

Astronomers and astrophysicists have always shared data and it is even truer today than ever before. The electronic data revolution for astrophysics and astronomy took real hold in the pre-Hubble spacecraft days with the establishment of the NASA/ADC (Astronomical Data Center) when catalogs listing data about a number of stars and other objects were digitized, recorded on CDs, and distributed at national and international meetings for free.

Although the least-squares method is the most powerful techniques that has been devised for the problems of data analysis, it is at the same time exceedingly critical. This is because the least-squares method suffers from the deficiency that, its estimation procedure does not have detecting and controlling techniques for the sensitivity of the solution to the optimization criterion of the variance $\tilde{\sigma}^2$ is minimum. As a result, there may exist a situation in which there are many significantly different solutions that reduce the variance $\tilde{\sigma}^2$ to an acceptable small value. At this stage we should point out that (1) the accuracy of the estimators and the accuracy of the fitted curve are to distinct problems; and (2) an accurate estimator will always produce small variance, but small variance does not guarantee an accurate estimator. This could be seen from property 3 (Section 3) of the least square solution by noting that the lower bounds for the average square distance between the exact and the least-squares values is $\tilde{\sigma}^2 / \lambda_{min}$ which may be large even if $\tilde{\sigma}^2$ is very small, depending on the magnitude of the minimum eigenvalue, λ_{min}, of the coefficient matrix of the least-squares normal equations.

Unless detecting and controlling the above mentioned situation, it is not possible to make a well-defined diction about the analysis of the observational data. Due to this difficulty and the importance of least squares for treating the nowadays incredible numbers of the available data, the present paper is devoted for the following goals : To develop an algorithm for model analysis of observational data in the sense of the least –squares criterion with error analysis.. By this algorithm one computes, all the solutions with their variances, the variance o the fit, the average square distance between the least square solution an the exact solution an graphical representation between the row and the fitted data. Mathematica module of the algorithm is also given.

OBSERVATIONAL DATA

Let $\{x_k\};k = 1,2,\ldots,m$, be a sequences of observational data points assumed free from errors. Corresponding to each x_k we have a number $\overline{f}_k$ of some function $f(x)$ at x_k which generally will be in error. Denote $f(x_k)$, the true value at x_k, by f_k and define:

$$\delta_k = \overline{f}_k - f_k \; ; \; k = 1,2,\ldots,m \tag{1}$$

as the observational errors at different data points.

The problem in the analysis of the observational data is to approximate or fit the data $\bar{f}_k$ by some function $\hat{p}(x)$ in such a way that $\hat{p}(x)$ contains or represents most (if not all) the information about $f(x)$ contained in the data and little (if any) of the errors. This is accomplished in practice by selecting a function

$$p(x) = p(x; c_1^{(n)}, c_2^{(n)}, \cdots, c_n^{(n)}), \tag{2}$$

which depends on the parameters $c_1^{(n)}, c_2^{(n)}, \cdots, c_n^{(n)}$. (The superscript n on $c_i^{(n)}$ denotes the fact that they will (generally depend on n). The function $p(x)$ may be linear or non-linear in the parameters $c_i^{(n)}$.

Since our study is the linear model analysis of the observational data, the function $p(x)$ will be selected as linear combinations of the parameters so that it takes the form:

$$p(x) = \sum_{i=1}^{n} c_i^{(n)} \Phi_i(x). \tag{3}$$

The $\{\Phi_i\}$ may, for example, be the set of monomials, exponentials, trigonometric polynomials, or indeed any arbitrary set of sufficiently defined functional values, provided only that they are linearly independent of the m values of x. Normally, n is small compared with the number, m of data points. In fact, the number n is unknown but for the most appropriate approximation to a given data the number n should be :

1-large enough so that the information about $f(x)$ in the data can well be represented by a proper choice of the parameters $c_i^{(n)}$, while at the same time n should be, 2-too small to avoid the fitting of the observed data too closely in the sense that the errors in the observed data are retained in the approximation.

Now the problem is to find the best estimate $\hat{p}$ of the function p i.e, to determine particular values $\hat{c}_1^{(n)}, \hat{c}_2^{(n)}, \ldots, \hat{c}_n^{(n)}$ for the parameters $c_i^{(n)}$ to obtain the best approximation:

$$\hat{p} = p(x, \hat{c}_1^{(n)}, \hat{c}_2^{(n)}, \cdots, \hat{c}_n^{(n)}) \tag{4}$$

Although there is agreement that the best estimate should give small deviations of the given data from the fitted curve- i.e., that the modulus of the quantity:

$$\varepsilon_k = p\left(x_k, c_1^{(n)}, c_2^{(n)}, \ldots\ldots, c_n^{(n)}\right) - \bar{f}_k \; ; \; k = 1,2,\ldots,m \tag{5}$$

should be small at each of the m points- there are differences of opinion as to how this to be achieved.

Let ε denote an error vector composed of the m deviations ε_k. Since x_k and $\bar{f}_k$ are regarded as fixed, the vector ε depends only on the parameters $c_1^{(n)}, c_2^{(n)}, \ldots, c_n^{(n)}$. Each common method for dealing with the competing requirements that ε_k are small, corresponds to the selection of a norm $\|\varepsilon\|$ for the vector ε. Some examples of norms in the real m- dimensional space are :

(i)
$$\|\varepsilon\|_1 \equiv \sum_{k=1}^{m} |\varepsilon_k| \tag{6}$$

(ii)
$$\|\varepsilon\|_2 \equiv \left(\sum_{k=1}^{m} \varepsilon_k^2\right)^{1/2} \tag{7}$$

(iii)
$$\|\varepsilon\|_\infty \equiv \max_{1 \le k \le m} |\varepsilon_k|. \tag{8}$$

The norm $\|\bullet\|_1$ is called Gershgorin norm, $\|\bullet\|_2$ the Euclidean norm and $\|\bullet\|_\infty$ the maximum normor, occasionally, the uniform norm. For any choice norm $\|\varepsilon\|$,the best estimate is one satisfying the requirement:

$$\| \boldsymbol{\varepsilon} \| = \text{minimum.} \tag{9}$$

Assuming one of the above norms, say the Q norm is chosen, then we can determine (by solving certain system of equations) the estimators $\hat{c}_i^{(n)}$ of the best estimate $\hat{p}$ of p that satisfies Equation (9), i.e., for the linear model analysis we have the best approximation:

$$\hat{p}(x) = \sum_{i=1}^{n} \hat{c}_i^{(n)} \Phi_i(x), \tag{10}$$

to be the true function $f(x)$ over $\{x_k\}$ in the sense of the Q norm fitting for the given $\Phi_i(x)$. Our best assumption is that for some unknown value of n, say N, the true function $f(x)$ can be expressed as a finite linear combination of the selected set of functions $\{\Phi_i\}$ in the sense of the Q–norm fitting ;that is, we assume

$$f(x)=\sum_{i=1}^{N}\hat{c}_i^{(n)}\Phi_i(x) \tag{11}$$

such that the Q norm of

$$\bar{f}-\sum_{i=1}^{n}\hat{c}_i^{(n)}\Phi_i(x_k)\equiv \text{minimum},\ \ k=1,2,\cdots,m;m\succ n \tag{12}$$

for some n=N and for a given $\Phi_i(x)$. Equation (11), with the condition of Equation (12), are the formulation of the linear model analysis of the observational data in the sense of Q-norm fitting.

Now it remains to decide what is the most efficient norm that should be used in the analysis of observational data. This decision constitutes very serious difficultly because:

1- The l norm $\|\varepsilon\|_l$ depends greatly on the distributions of the observational errors δ_k.

2- There is a large variation in the effectiveness of various norms and no single norm is good (or even mediocre).

3- The distributions of the observational errors cannot be determined with any precision.

It is undoubtedly true that the Euclidean norm is the most efficient when δ_k. are normally distributed, and in this case the estimators are the maximum likelihood estimators. It is natural therefore, to consider in our study the assumption that δ_k.are normally distributed and analyzed the linear –model representation of the observational data in the sense of the Euclidean norm-i.e., in the least –squares criterion.

Before starting the analysis, however, we introduce additional conditions concerning the character of the errors δ_k and assume that $\delta_1,\delta_2,\cdots,\delta_k$ are unbiased (without systematic error),independent and normally distributed with zero mean and variance $\tilde{\sigma}^2$. These conditions indicate that:

$$E(\delta_k)=0, E(\delta_p\delta_q)=E(\delta_p)E(\delta_q);$$

$$\delta_k\,\varepsilon\, N(0,\tilde{\sigma});\sigma_k^2=E(\delta_k^2)=\tilde{\sigma}^2$$

E(z) being the expectations of z and $\tilde{\sigma}^2$ the unknown variance.

In the case of weighted observations, Equation (16) is replaced by:

$$\sigma_k^2 = E(\delta_k^2) = \frac{\tilde{\sigma}^2}{\omega_k}, \tag{13}$$

where ω_k are the weights of the observations $(\omega_k = \omega(x_k))$. Therefore the problem of this section is the application of the least-squares method for the linear model representation of the observational data.

LEAST-SQUARES APPROXIMATION

Below is a list of vectors and matrices used in the present section:

Symbol	Order	Element
$\bar{\mathbf{F}}$	$m\times 1$	$\bar{f}_k$
$\mathbf{F}$	$m\times 1$	f_k
$\boldsymbol{\delta}$	$m\times 1$	δ_k
$\hat{\mathbf{C}}$	$n\times 1$	$\hat{c}_i^{(n)}$
$\mathbf{C}$	$n\times 1$	$c_i^{(n)}$
$\boldsymbol{\Phi}$	$n\times m$	$\Phi_{ik} = \Phi_i(x_k)$
$\boldsymbol{\omega}$	$m\times m$	$\omega_{kk} = \omega_k$; $\omega_{ql} = 0; q \neq l.$

The transpose of a vector or a matrix will be indicated by the superscript 'T'.

Derivation of the Normal Equations

In matrix notation, the problem of the weighted least-squares approximation is to determine the estimators $\hat{c}$ such that

$$e_n(\hat{\mathbf{C}}) = \mathbf{v}^T \boldsymbol{\omega}\, \mathbf{v} = \text{minimum}, \tag{14}$$

where the vector of the residuals $\mathbf{v}$ is defined as

$$\mathbf{v} = \bar{\mathbf{F}} - \boldsymbol{\Phi}^T \hat{\mathbf{C}}. \tag{15}$$

From Equations (14) and (15) we have ;

$$e_n(\hat{C}) = (\bar{F} - \Phi^T\hat{C})^T \omega \ (\bar{F} - \Phi^T\hat{C})$$
$$= (\bar{F}^T - \hat{C}^T\Phi)\omega \ (\bar{F} - \Phi^T\hat{C})$$
$$= (\bar{F}^T\omega - \hat{C}^T\Phi\omega) \ (\bar{F} - \Phi^T\hat{C})$$

that is

$$e_n(\hat{C}) = \bar{F}^T\omega\bar{F} - \hat{C}^T\Phi\omega\bar{F} - \bar{F}^T\omega\Phi^T\hat{C} + \hat{C}^T\Phi\omega\Phi^T\hat{C}. \tag{16}$$

Since $\bar{F}^T\omega\Phi^T\hat{C}$ is scalar then,

$$\bar{F}^T\omega\Phi^T\hat{C} = (\bar{F}^T\omega\Phi^T\hat{C})^T = [(\Phi^T\hat{C})^T.(\bar{F}^T\omega)^T]$$

that is:

$$\bar{F}^T\omega\Phi^T\hat{C} = \hat{C}^T\Phi\omega^T\bar{F}, \tag{17}$$

Since

$$\omega^T = \omega, \tag{18}$$

then

$$\bar{F}^T\omega\Phi^T\hat{C} = \hat{C}^T\Phi\omega\bar{F} \tag{19}$$

From Equations (16) and (19) we get ;

$$e_n(\hat{C}) = \bar{F}^T\omega\bar{F} - 2\hat{C}^T\Phi\omega\bar{F} + \hat{C}^T\Phi\omega\Phi^T\hat{C}. \tag{20}$$

Equation (20) could be written as:

$$e_n(\hat{C}) = \text{constant} - 2\hat{C}^T\mathbf{b} + \hat{C}^T G \ \hat{C}. \tag{21}$$

$$\left.\begin{aligned} \text{constant} &\equiv \bar{F}^T\omega\bar{F} \\ \mathbf{b} &\equiv \Phi\omega\bar{F} \\ G &\equiv \Phi\omega\Phi^T \end{aligned}\right\} \tag{22}$$

The necessary conditions for the minimum are

$$\frac{\partial e_n(\hat{\mathbf{C}})}{\partial \hat{\mathbf{C}}} = \mathbf{0}.$$

Consequently, from Equation (21) we get

$$-2\mathbf{b} + 2\mathbf{G}\hat{\mathbf{C}} = \mathbf{0}, \tag{23}$$

therefore

$$\mathbf{G}\hat{\mathbf{C}} = \mathbf{b} \tag{24}$$

where the general elements are given by

$$g_{ij} = \sum_{k=1}^{m} \omega_k \Phi_{ik} \Phi_{jk} \tag{25}$$

and

$$b_i = \sum_{k=1}^{m} \omega_k \Phi_{ik} \overline{f}_k. \tag{26}$$

Equations (24) are called the normal equations of the weighted least-squares approximations for the linear –model representations of the observational data. These equations represent a set of n linear equations in n unknowns $\hat{c}_i^{(n)}; i = 1, 2, \cdots, n$ .The coefficient matrix G is symmetric and is positive definite (i.e., all eigen values $\lambda_i; i = 1, 2, \cdots, n$ are positive) if all the fundamental functions $\Phi_i(x)$ are linearly independent for the arguments x_k ; that is, if the rows of Φ are linearly independent. It should be noted that the normal Equations (24) represent a relation between two exact quantities in the sense that if the matrix G and the right-hand side b represent the observational data exactly, then there is a vector $\hat{C}$ which satisfies exactly the least –squares criterion. Of course, in practice such an exact situation no longer exists. In the following section, some important properties of $\hat{C}$ (in the sense described above) will be given.

Some Properties of $\hat{C}$

Property 1

1-The estimators $\hat{C}$ by the method of least-squares gives the minimum of e_n

Property 2

2- The estimators $\hat{C}$ by the parameters C, obtained by the method of least-squares are unbiased; i.e.

$$E(\hat{C}) = C.$$

Property 3

3-The variance –covariance matrix $Var(\hat{C})$ *of the unbiased estimators $\hat{C}$ is given by*

$$Var(\hat{C}) = \tilde{\sigma}^2 G^{-1}.$$

Property 4

4-The average squared distance between $\hat{C}$ and C is

$$E(L^2) = \tilde{\sigma}^2 \sum_{i=1}^{n} \frac{1}{\lambda_i}$$

Property 5

5-The average squared Euclidean norm of $\hat{C}$ is

$$E(\hat{C}^T \hat{C}) = C^T C + \tilde{\sigma}^2 \sum_{i=1}^{n} \frac{1}{\lambda_i}$$

According to the conditions of Equations (13) and (17) it was proved [1] the following properties of $\hat{C}$:

Also it should be noted that:

® If H(u) is a function of a measured quantity u, then the standard error $\tilde{\sigma}_H$ then

$$\tilde{\sigma}_H = \frac{dH}{du} \tilde{\sigma}_u \tag{27}$$

where $\tilde{\sigma}_u$ is the standard error of u

® If the precision is measured by the probable error e, then

$$e = 0.6745\ \tilde{\sigma}. \tag{28}$$

MATHEMATICA SIMULATION

Mathematica Module: Linear Model

• **Purpose**

To fit observational data (x,y)(say) to the linear model $y = \sum c_i \Phi_i(x)$ in the sense of least-squares criterion with its error analysis.

• **Input**

x1: List of the independet varibles of the data.

y1 : List of the dependet varibles of the data.

f : List of the basic functions used for the linear model representation of the data.

I1 ,I2 ,I3 : Positive numbers each $\in[0,1]$ used to disply color drawing according to the built in function

Backgroung → RGBColor[I1,I2,I3].

ind : A charcter indecating the name of the independent variable, (x for example).

dep: A charcter indecating the name of the dependent variable, (y for example).

• **Output**

The c's coefficients and their probable errors.

The probable error of the fit.

The average square distance between the exact and the least squares solutions.

Graphical representations

• **Needed procedures**

None

• List of the Module

```
LinearModel [x1_List,y1_List,I1_,I2_,I3_,ind_,dep_] :=
Module[{str, stt, n, m, b, eO, ma, mai, qf, pro, dcs, ei,qv, es, sx, sy, sx2, sy2,sxy, rc}
( str=Min[x1];stt=Max[x1];n=Length[f];m= Length[xl];
Print[ StyleForm["Liner Least –Squares Fit by the Model: ",FontWeight->"Bold",
FontSize->12],
StyleForm[dep, " ",fontWeight->"Bold",FontSize-> 12],
StyleForm[=, " ",FontWeight->"Bold",FontSize-> 12],
StyleForm[" ∑_{i=1}^{n} ci Φi,Where ",fontWeight->"Bold",FontSize-> 12],
StyleForm[f[[i]]/ξ-> ind//TraditionalForm, FontWeight->"Bold",FontSize->12]], {i,1,n}],
C [l_,k_] := f [[ l ]] /. ξ -> xl [{ k ]] ;
B = Table [ Sum [ C [ i, k ] * yl[[k]], {k, m}], {i, 1,n}]; e0=Sum[y1[[k]] ^2, {k, m}];
ma= Table[Which[i<=j,g[I,j] = ∑_{K=1}^{m} c[ i , k] * c[j, k], i > j ,
g{i,j]=g[j,i]] ,{i,1,n}, {j,1,n}] ; mai=Inverse[ma]; cs=mai.b; qf=Expand[cs.ma.cs];
sigma=Sqrt(e0-qf)/ (m-n)]; pro=0.6745 * sigma;dcs=pro * Table[sqrt[mai[[i,i]]], {i,1,n}];
ei=Eigenvalues[N[ma]]; qv=sigma ^2 * Sum[1./ei[[j]], {j,n}]; es=qv+Sum[cs[[k]]^2,{k,n}];
funFit= ∑_{i=1}^{n} cs [[ i ]] * f[[i]];sx=Apply[plus,x1];sy= Apply[plus,y1] ; sx2 Apply[plus, x1^2];
sy2= Apply[Plus,y1^2]; sxy= Apply[plus, x1*y1];
```

```
Print[ StyleForm["(2) The Solutions and their probable Errors" ,FontWeight->"Bold",
FontSize-> Do[Print[StyleForm["="ci,FontWeight->"Bold",FontSize->12],
StyleForm["=",FontWeight->"Bold", FontSize-> 12],
StyleForm[cs[[i]],FontWeight->"Bold", FontSize->12],StyleForm["±" ,FontWeight->"Bold", FontSize
  StyleForm[dcs[[i]],FontWeight->"Bold", FontSize->12]],{i,l,n}],
Print[ StyleForm["(3) The Probable Error of the Fit is" ,FontWeight->"Bold", FontSize->12]];
Print[ StyleForm[" "e,FontWeight-> Bold", FontSize->12],
StyleForm["=", FontWeight->"Bold", FontSize->12],
```

```
StyleForm[pro,fontWeight->"Bold", FontSize->12]];
Print[ StyleForm["(4)The Average Squared Distance Between ĉ and c is" ,
  FontWeight->"Bold",fontSize-> 12]];
Print[StyleForm[" "Q, FontWeight->"Bold",FontSize->
12], StyleForm["=",FontWeight->"Bold",FontSize->12],
StyleForm[qv,FontWeight->"Bold",FontSize->12]] ;
Print[ styleForm["(5) Graphical Representations" ,FontWeight->"Bold,FontSize->12]];
Print[ StyleForm[" 5-a GraphoftheRawData" ,FontWeight-> "Bold",FontSize->12]];
datacorr=Table[{x1[[i]],yl[[i]]}, {i,1,m}];dataPlot=ListPlot[datacorr, PlotStyle-> PointSize[O.
Background→RGBColor[I1, I2,I3],AxesLabel→ {"x","y"}];
 Print[ StyleForm["  5-b GraphoftheRaw and the Fitted Data " ,
  FontWeight-> "Bold",FontSize->I2]];
fitPlot=Plot[funFit/. ξ-> τ , {τ, str, stt}, Plotable →"v",DisplayFunction ->
  Identity];
Show[{dataPlot, fitPlot}, DisplayFunction -> $DisplayFunction]; ) ]
```

We applied this module for many problems for galactic structure, the results are very satisfactory and will be published soon in astronomical journals

In concluding the present paper, an algorithm for model analysis of observational data was deloped in the sense of the least –squares criterion with full error analysis.. By this algorithm one computes, all the solutions with their variances, the variance of the fit, the average square distance between the least square solution and the exact solution, and graphical representation between the row and the fitted data. Mathematica module of the algorithm was established, through five points, its purpose –input - output - needed procedures and the list of the module. By this paper we have been tried to produce an error controlled algorithm of the least squares method for observational data.

REFERENCES

1. Kolb, E.W., and Turner, M.S. 1990, The Early Universe. Frontiers in Physics. Redwood City, CA: Addison –Wesley. Advanced Book Program.
2. Kopal, Z and Sharaf, M.A. 1980, Linear Analysis of the Light Curves of Eclipsing Variables, Astrophysics and Space Science, 70, 77-101.
3. Borne, J, M., and 91 others: 2009, Astroinformatics: A 21st Century Approach to Astronomy. Tech. rept. Submitted to 2010 Decadal Survey.

4

ON THE CONSTRUCTION OF FAMILIES OF TYPE II_1 SUBFACTORS EACH CONTAINING A MIDDLE SUBFACTORS

Bahman Mashood*, Mahmood Khoshkam[1]

[1*] Former lecturer at university of Saskatchewan, San Francisco/California

ABSTRACT

In this article we are going to construct a family of type Π_1 subfactors each containing a middle subfactor. As a result of the above construction we show that the set of the indices of hyperfinite irreducible subfactors contains the interval [37.0037,∞).

1. INTRODUCTION AND PRILIMINARIES

In the next chapter we are going to use locally trivial subfactors to construct a set of middle subfactors. The important of locally trivial subfactos was indicated by S.Popa at [9]. The simplest locally trivial subfactors are those having only two orthogonal projections in their relative commutant. It is well known that these subfactors are isomorphic to Jones subfactors.. Also it is easy easy to show that these kind of subfactors do not posses middle subfactors. For the locally trivial subfactors that their relative commutant have dimension larger than two we believe that the above result is still valid ie'there are no middle subfactors. Suppose we are Given a pair $L \subset M$ of subfactors that are limiting algebras of a tower of commuting squares.

Then using the results in [1] we can show that if the inclusion graphs of the corresponding finite C* algebras are A graphs, then the subfactors $L \subset M$ do not have middle subfactors. One of the problems in the index theory, is to find the set of all the values for the indices of hyperfinite irreducible subfactors. Using the above constructions, we are going to show that the above set contains the interval (37.0037,∞).

2. MAIN RESULTS

For a given pair of subfactors $N \subset M$, With $[M:N] = \lambda^{-1} < \infty$. Let e be a projection in M that induces the expectation N onto $Q = (e)' \cap N$. Let $P_1, P_2 = 1 - P_1$ be a partition of unity in Q. Using standard arguments as in there exists an isomorphism Φ, taking NP_1 onto NP_2.. Let $L(P_1)$ be the set of all the elements of the form $L(P_1) = (x + \Phi(x); x \in NP_1)$.. Then it is well known that $L(P_1)$, is a locally trivial subfactor of N. Also by (Lemma2.2.1) [5], $[N : L_P] = 1 = tr(P_1) + 1 = tr(P_2)$. Where tr is a unique normalized trace on M. Suppose P_1, does not communicate with $e_0 = e$. Set $y = P_1 e P_2 \neq 0$. Note that the relative commutant of $L(P_1)$, inside M is spanned by the projection P_1. Since y, does not communicate with P_1, we have that the algebra, $H(P_1) = < L(P_1); Y >$ is a middle subfactor which is strictly larger than $L(P_1)$ If under certain conditions $H(P_1)$ becomes strictly smaller than M, then $H(P_1)$ becomes a proper middle subfactor. In this case by the above arguments it is easy to see that the inclusions $H(P_1) \supset L(P_1)$ and $H(P_1) \subset M$ are irreducible inclusion of subfactors. Let us denote $r_1 = [H(P_1) : L(P_1)]$ and $r_2 = [M : H(P_1)]$. Let IR, be the set of all irreducible subfactors of finite index. Let us denote by IIR, the set of indices of all subfactors in IR. Then by (Proposition2.1.15) [5] r1r2 is in IIR. It is easy to check that the set of of the elements f of the form f=∑j∈J ujgzjegwj, with $g = E_Q(P_2)$, $z_j \in Q$, $u_j = P_1 x_j P_1$, $w_j = P_1 y_j P_1$, $x_j \in N$ $y_j \in N$ where J is a set of indices is dense in (H(P1))P1. Let e1 be a projection in N, such that $E_Q(e_1) = \lambda$. Then e1 induces the expectation of Q onto the subfactor Q1 = (e)′∩Q. Next for a number r, r∈IIR, construct an irreducible subfactor Q_2, $Q_2 \subseteq Q_1$, with [Q1 : Q2] = r. We can define the projection e3, Using (corollary 1.8) [13], there exits a projection e3 in Q1, such that e3 induces the expectation of Q2 onto The subfactor Q3, with Q3 = (e3)′∩Q2. This process will induce the following tonnel, $M \supset N \supset Q \supset Q_1 \supset Q_2 \supset Q_3$. Let us set P1 = qe1e3 with q a projection in Q3. Now we can check that the following set of elements, f of the form, $f = P_1(1-\lambda)^2 e_0\left(\sum_{j\in J} x_j z_j y_j\right)P_1 = (1-\lambda)^2 P_1 e_0 z P_1$, with $z = \sum_{j\in J} x_j z_j y_j$, with x_j, z_j, y_j as in the above and J a set of indices will be a dense subset of (H(P1))P1. In particular assuming now that H(P1) = M implies that the above set of elements are dense in MP1. Furthermore as we mentioned in the above for any number r∈IIR; e3∈Q1 can be chosen such that tr(e3) = r. For example suppose tr(e1) = tr(e0) = .5. Then it is easy to see that there exists a unitary $V \in (e_3)' \cap N = L$, with L a type π_1 Von Neumann algebra, such that V e1V* = 1 − e1. Then we can express the f from the above as f = λ(1 −λ)2e1e3ze1e3 + (e1e3e0V e1e3)(e1e3V*ze1e3). For a given real number S, let [S], be the largest integer which is smaller or equal to [S]. Let us set Sr = S − [S]. Let us assume now that H(P1) = M, We will get the following results.

Lemma 1 Keeping the same notations as in the above let $\lambda^{-1} = \left[\lambda^{-1}\right] + \lambda_r^{-1}$ Then there exist unitary operators U, U_2, U_3 in L and projection $p \leq e_1$, such that f can be expresses as in the following, $f = \left(1-\lambda^2\right)\left(e_1 e_3 e_0 q e_1 q^* z e_1 e_3\right.$ $+ e_1 e_3 e_0 U e_1 q^* U^* z e_1 e_3 + \sigma_1 e_1 e_3 e_0 U_2 e_1 U_2^* z e_1 e_3$ With σ1, will be equal to 1 if

$\left[\lambda^{-1}\right]$ odd integer and equal to zero otherwise. Similarly σ2 is equal to 1 if λ^{-1} is not an integer and equal to zero otherwise.

Proof First suppose $tr(e_0)^{-1} = \lambda^{-1} = 2n$, for some positive integer n, then let $f_1 = e_1, f_2, \ldots\ldots, f_{2n}$ be a partition of unity by orthogonal projections in L, such that for any odd integer $k < 2n$, there exists a unitary $V_k \in L_{f_k + f_{k+1}}$, with $f_k = V_k f_{k+1} V_k^*$. Using our definition of $f_1 = e_1$, this implies,

$$f = (1-\lambda)^2 \left(\sum_{i=1}^{n} \begin{pmatrix} e_1 e_3 e_0 f_{2i-1} z e_1 e_3 \\ + e_1 e_3 e_0 V_i f_{2i-1} V_i^* z e_1 e_3 \end{pmatrix} \right)$$

Let $U = \sum_{i=1}^{i=n} V_i$ Then U is a unitary in L. Set $g = \sum_{i=1}^{i=n} f_{2i-1}$ The we have, $f = (1-\lambda)^2 \left(e_1 e_3 e_0 g z e_1 e_3 + e_1 e_3 e_0 U g U^* z e_1 e_3 \right)$ Next for each $1 \le i \le n$, there exists a unitary $m_i \in L_{e_1 + f_{2i-1}}$ such that $m_i e_i m_i^* = f_{2i-1}$ Hence $g = \sum_{i=1}^{i=n} m_i e_i m_i^*$ Next since for $i \neq j$, $\left(m_i e_i m_i^* \right)\left(m_j e_j m_j^* \right) = 0$. Thus we get $tr\left(m_i e_1 m_i^* m_j e_1 m_j^* \right) = 0$. For $i \neq j$, let us define $y = e_1 m_j^* m_i e_1$. Then it is easy to check $tr\left(y y^* \right) = 0$. This implies that y = 0. Another useful relation that we will need later is the following equality, $e_1 q q^* e_1 = n e_1$ that can be checked easily. Let us define the operator h, with $h = \left(\sum_i m_i \right) e_1 \left(\sum_k m_k^* \right) = q e_1 q^*$ with, $1 \le i, k \le n$ and $q = \sum_i m_i$. Then using the above relations we can see that h is a projection, tr(h) = tr(g), and UhU^* is orthogonal toh. Hence we get $tr(h) = tr(UhU^*) = .5$. This will implies that f can be expressed as, $f = (1-\lambda)^2 \left(e_1 e_3 e_0 q e_1 q^* z e_1 e_3 + e_1 e_3 e_3 U q e_1 q^* U^* z e_1 e_3 \right)$ Suppose λ−1 = 2n+1. Then we have the following partition of unity, $e = f_1, f_2, \ldots\ldots, f_{2n}, f_{2n+1}$. Where the above projections have equal traces. Furthermore ther exists a unitary $U_2 \in L_{e_1 + f_{2n+1}}$ such that f can be expressed as,

$$f = (1-\lambda)^2 \begin{pmatrix} e_1 e_3 e_0 q e_1 q^* z e_1 e_3 + e_1 e_3 e_0 U q e_1 q^* U^* z e_1 e_3 \\ + e_1 e_3 e_0 U_2 q e_1 q^* U_2^* z e_1 e_3 \end{pmatrix}.$$

then we have the following partition of unity by the following projections, $e = f_1, f_2, f_3, \ldots\ldots, f_{2n}, \sigma_1 f_{2n+1}, \sigma_2 f_0$. Where σ1 and σ2, can only take values 0 or 1, depending if the corresponding projections f2n+1 and f0 are or are not equal to zero. Furthermore for k ≠ 0, all non zero projections fk's, have equal traces and tr(f0) < λ. Hence generally f can be expressed as,

$$f=(1-\lambda)^2\begin{pmatrix} e_1e_3e_0qe_1q^*ze_1e_3 \\ +e_1e_3e_0Uqe_1q^*U^*ze_1e_3 \\ +\sigma_1e_1e_3e_0U_2e_1U_2^*ze_1e_3 \\ +\sigma_2e_1e_3e_0U_3e_1pU_3^*ze_1e_3 \end{pmatrix}.$$

Where U and U2 are as in the above, U3 a unitary in $L_{e_1+f_0}$ and p, is a sub projection of e1 and is in L_{e_1+f}. At this point note that we can extend U2 and U3 to be unitaries in L. Finally we can express f, as

$$f=(1-\lambda)^2\begin{pmatrix} (e_1e_3e_0qe_1e_3)(e_1e_3q^*ze_1e_3) \\ +(e_1e_3e_0Uqe_1e_3)(e_1e_3q^*U^*ze_1e_3) \\ +\sigma_1(e_1e_3e_0U_2e_1e_3)(e_1e_3U_2^*ze_1e_3) \\ +\sigma_2(e_1e_3e_0U_3e_1e_3p)(e_1e_3U_3^*ze_1e_3) \end{pmatrix}.$$

Now let us set the following notations. $n_1=e_1e_3e_0qe_1e_3$, $n_2=e_1e_3e_0Uqe_1e_3$, $n_3=e_1e_3e_0U_2e_1e_3$, $n_4=e_1e_3e_0U_3e_1e_3$ and $G=N_{e_1,e_3}$.

Lemma 2 Keeping the same notations as in the above, and assuming that σ_1 and σ_2 both different from zero and without loss of generality, we have the following equalities.

$$E_{N_{e_1,e_3}}\left(n_k^*n_k\right)=e_1e_3, j\neq k, E_{N_{e_1,e_3}}\left(n_j^*n_k\right)=0, k=1,1,3,4.$$

Proof Note that since e_3 commues with all the above operators, dropping e_3 from all the operations does not makes any different from the final outcome. So in the following operations we ignore the existence of e_3. In particular we can identify N_{e_1} with $G=N_{e_1}$. Since the proof of the above equalities are very similar, we only show some of the equalities. We have $n_1n_1^*=e_1e_0qe_1q^*e_0e_1$. Hence we get, $E_G\left(n_1n_1^*\right)=\lambda^2e_1qq^*e_1=n\lambda^2e_1$. Next $n_2^*n_1=e_1q^*U^*e_0e_1e_0qe_1=\lambda e_1q^*U^*e_0qe_1$. Thus $E_G\left(n_2^*n_1\right)=\lambda^2e_1q^*U^*qe_1=\lambda^2e_1q^*U^*qe_1$. But we have, $e_1e_1qq^*e_1/n=e_1qq^*e_1/n=e_1$. This implies, $E_G\left(n_2^*n_1\right)=n^{-2}\lambda^2e_1q^*qe_1q^*U^*qe_1q^*qe_1=n^{-2}\lambda^2e_1q^*U^*U\left(qe_1q^*\right)U^*\left(qe_1q^*\right)qe_1$. But $U\left(qe_1q^*\right)U^*$ is orthogonal qe_1q^* which implies that $E_G\left(n_2^*n_1\right)=0$. Furthermore $n_3^*n_3=e_1U_2^*e_0e_1e_1e_0U_2e_1=\lambda e_1U_2^*e_0qe_1$. Hence we get, $E_G\left(n_3n_3^*\right)=\lambda^2e_1$. Next $n_3^*n_1=e_1U_2^*e_0e_1e_1e_0qe_1$ $=\lambda e_1U_2^*e_0qe_1$. This implies, $E_G(n_3n_1)=\lambda^2e_1U_2^*qe_1=e_1U_2^*qe_1q^*qe_1/n=U_2^*U_2e_1U_2^*q^*e_1q^*qe_1/n$. But using the above relations, $\left(U_2e_1U_2^*\right)\left(qe_1q^*\right)=0$, which implies $E_G\left(n_3^*n_1\right)=0$. Now let us calculate $E_G(n_1^* n_4)$. $E_G(n_1^* n_4) = E_G(e_1 q^* e_0 e_1 e_1 e_0 U_3 e_1 p) = \lambda^2 e_1 q^* U_3 e_1 p = \lambda^2 e_1 q^* U_3 e_1 p U_3^* U_3$. Now using the above relations we can show that $e_1 q^* U_3 p U_3^* = 0$, hence $E_G(n_1^* n_4) = 0$.

Assuming that Me1e3 is acting standardly on H = [Me1e3] and L(P1) = M, the above lemma implies that the operator identity can be spanned by at most four orthogonal projections in G each of trace less or equal to λ. Hence by Remarks(1.4) [13], we get the following Corollary,

Corollary 3 Keeping the same notations as before, for [M:N] > 4, $H(P_1)$ is a proper middle subfactor.

As before let IIR represents the set of all indices of irreducible hyperfinite subfactors.

Suppose H(P1) is a proper middle subfactor, ie; L(P1) ⊂ H(P1) ⊂ M, where the inclusions are restrict. Let us denote r1 = [H(P1):L(P1)] and r2 = [M:L(P1)]. Also using the fundamental property of the index of subfactors, r1r2 = r = [M:L(P1)] = [M : N][N : L(P1)]. But [M : N] =λ−1 and [N:L(P1)] = (tr(P1)(tr(P2))−1 Hence r = r1r2 = λ−1(tr(P1)tr(P2))−1 And by the results of [5], r∈IIR. Now notice that by the result of S.Popa in "Subfactors and classification in von Neumann algebras" Corollary(4.4) of the above article indicate the gap in IIR between the values 4 and $2+\sqrt{5}$. In fact $2+\sqrt{5} \simeq 4.026$, corresponds to the square of the norm of Coxeter graph E10 and there exists a subfactor of such an index. By its definition P1 = qe1e3 and we had λ = tr(e1) λ1 = tr(e3). Hence tr(P1) = tr(q)λλ1. Let us set c = tr(q), and $\varpi = 2+\sqrt{5}$ then c can take any value in the interval [0,1]. Note also that (λ1)−1can take any value in IIR larger than .5. Let us denote, $\alpha = \lambda\lambda_1$, $\beta = c\alpha$, then β = tr(P1). Note that $(\alpha)^{-1}$ can take any value in IIR larger than or equal to 2ϖ. Hence if we set $\beta = c\lambda\lambda_1$, then we have $0 \le \beta \le 1/(2\varpi)$. Since q can be taken to be any projection in Q3, tr(P1), can get any value in the interval $[0,1/(2\varpi)]$. This implies the following theorem.

Theorem 4 Keeping the same notations as in the above, suppose $\lambda^{-1} = \varpi$ and $(\lambda_1)^{-1} = 2$. Let $N \supset M$, $[M:N]=\lambda^{-1}$ be a pair of irreducible subfactors and lets define a projection $p = qe_1e_3$, for some projection q in Q_3. Then $H(P_1) = \langle L(P_1), p_1 e p_2 \rangle$ is a proper irreducible subfactor of M. In particular letting q to vary in Q_3, we will get that *IIR* includes the interval $[\varpi/(1/2\varpi)(1-1/2\varpi), \infty] \simeq [37.0037, \infty)$.

At this end note that by the above arguments there exists a function Φ, acting on the above interval

REFERENCES

1. D.Bisch. A note on intermediate subfactors. Pac.J.Math.163(1994), 201-216.
2. E.Christensen, Subalgebras of finite algebra, Math.Ann.243(1979), 17-29.
3. F. Goodman, P. S. Delaharpe and V. Jones, Coxter graph and towers of algebras. Math. Sci. Res. Hist.Publ., Springer-Verlag, V. 14, 1989.

4. U.Haagerup, Principal graphs of a subfactors, Proceeding of the Tanguchi Simposium on operator algebras, World scientific, 1999.
5. V.Jones, Index for subfactors, Invent Math 72 (1983), 1-25.
6. V.Jones, V.S.Sunder, Introduction to Subfactors. Combridge University Press. 1997.
7. M.Khoshkam and B.Mashood, Rigidity in commuting squres, New Zeland Journal Of Math. V. 32(2003), 37-56.
8. M.Khoshkam and B.Mashood, On the relative commutants of subfactors, Proceeding of A.M.S 126(1998)2725-2732.
9. B.Mashood, On a ladder of commuting squares and a pair of factors generated by them, preprint(1990) Dept. Math. University of Victoria. Canada. BC.V8W 2Y2.
10. B.Mashood, On the tower of relative commutants part-I, Math Japonica 37, No.5(1992)901-906.
11. B.Mashood, and K.Taylor, On the continuity of the index of subfactor of a finite factor, J.Func.anal.76 no.11(1988)56-66.
12. A.Ocneanu, Quantized groups, string algreba, operator algebra and application, London Math Society. Lecture notes ser. 136 119-172, Cambridge University Press, 1988.
13. M.Pimsner, S.Popa, Entropy and Index for subfactors., Ann.Sci.Ecole Norm. Sup. 19 (1986).
14. M.Pimsner, S.Popa, Finite dimensional approximation for pair of algebras and obstruction for index, J. Func. Anal. 98(1991), 270-291.
15. M.Pimsner, S.Popa, Sur les sous facteurs d'indice fini d'facteur de type II ayant propriet T,C.R.Acad.Sci, 303(1986), 359-361.
16. S.Popa, Classifcation of amenable subfactors of type II., Acta Math., 172(1994), 163-255.
17. S.Popa, Sur la classification d' facteur hyperfini, Comtes rendus (1990), 95-100.
18. N.Sato, Two subfactors arising from a non-degenerate commuting squares, Pac.J.Math.Vol.180, No 2, 1997, 369-376.
19. H.Wenzl, Hecke algebras of type An and subfactors, Invent, Math. 92(1998) 349-383

5

SIMILAR CONSTRUCTING METHOD FOR SOLVING THE BOUNDARY VALUE PROBLEM OF THE COMPOUND KUMMER EQUATION

Duo Zhang[1], Shunchu Li[1],, Dongdong Gui[2]

[1] College of Science, Xihua University, Chengdu, China

[2] Beijing Dongrunke Petroleum Technology Co., Ltd., Beijing, China

ABSTRACT

This article is devoted to analyze the compound Kummer equation. Through the structure of left region and right region similarity kernel function, have similar structure of solution for this boundary value problems. The left field solution of this kind of boundary value problems can be obtained by combing the coefficients of boundary value problems on the left and the left area similar kernel function. The right field solution of this kind of boundary value problems can be expressed by assembling the left, the right area similar kernel function, the left lead solution functions and the coefficients of the convergence conditions. Thus put forward the simple method which was named similar constructive method to solve the boundary value problems of the compound Kummer equation. The put forward of this method provides the convenience for solving this kind of differential equation boundary value problems.

INTRODUCTION

The boundary value problems of differential equations always are needed in the calculation of many physical and engineering [1,2,3,4,5]. So it is not just the initial value problems of differential equation, boundary value problems are also the focus of our study. Compound boundary value problem of Kummer equation is a mathematical problem that closely associated with the practical application. But when in solution of the compound boundary value problem of Kummer equation, we often encounter the complicated solving processe. So it is inconvenience for the practical application. Some boundary value problems of differential equations can be written in the form of solution of continued fractions which was named similar structure type. And this conclusion in 2004 has been proved [6,7,8,9,10]. For example, the boundary value problem of Kummer equation, the boundary value problem of Euler hypergeometric equation. Then the solution of the boundary value problem for composite Kummer equations whether it has the above properties?

This article will study the following compound confluence hypergeometric equation boundary value problem:

$$\begin{cases} xy_1''+(\gamma_1-x)y_1'-\alpha_1 y_1=0(a<x<c) \\ xy_2''+(\gamma_2-x)y_2'-\alpha_2 y_2=0(c<x<b) \\ [Ey_1+(1+EF)y_1']_{x=a}=D \\ y_1|_{x=c}=my_2|_{x=c} \\ y'_1|_{x=c}=ny'_2|_{x=c} \\ [Gy_2+Hy'_2]_{x=b}=0 \end{cases} \tag{1}$$

Where $\alpha_1,\alpha_2,\gamma_1,\gamma_2,D,E,R,G,H,a,b,c,m,n$ are all known real constant, and γ_1,γ_2 are not as integers, $D\neq 0, G^2+H^2\neq 0, a<c<b$.

PRELIMINARY KNOWLEDGE

lemma 1 [11] The general solution for $xy''+(\gamma-x)y'-\alpha y=0$ is

$$\begin{aligned} y= &AF(\alpha,\gamma,x) \\ &+Bx^{1-\gamma}F(\alpha+1-\gamma,2-\gamma,x) \end{aligned} \tag{2}$$

Where γ is not an integer, both A and B are real constants and

$$F(\alpha,\gamma,x)=\sum_{k=0}^{\infty}\frac{(\alpha)_k}{k!(\gamma)_k}x^k$$

lemma 2 Introducing two binary functions

$$\begin{aligned} &\phi_{0,0}^{i}(x,\xi) \\ &=\xi^{1-\gamma_i}F(\alpha_i,\gamma_i,x)F(\alpha_i+1-\gamma_i,2-\gamma_i,\xi) \\ &-x^{1-\gamma_i}F(\alpha_i+1-\gamma_i,2-\gamma_i,x)F(\alpha_i,\gamma_i,\xi) \\ &\qquad (i=1,2) \end{aligned} \tag{3}$$

Then

$$
\begin{aligned}
\phi_{1,0}^{i}(x,\xi) &= \frac{\partial \phi_{0,0}^{i}(x,\xi)}{\partial x} \\
&= \frac{\alpha_i}{\gamma_i}\xi^{1-\gamma_i}F(\alpha_i+1,\gamma_i+1,x)F(\alpha_i+1-\gamma_i,2-\gamma_i,\xi) \\
&\quad -(1-\gamma_i)x^{-\gamma_i}F(\alpha_i+1-\gamma_i,2-\gamma_i,x)F(\alpha_i,\gamma_i,\xi) \\
&\quad -\frac{\alpha_i+1-\gamma_i}{2-\gamma_i}x^{1-\gamma_i}F(\alpha_i+2-\gamma_i,3-\gamma_i,x)F(\alpha_i,\gamma_i,\xi)
\end{aligned}
\tag{4}
$$

$$
\begin{aligned}
\phi_{0,1}^{i}(x,\xi) &= \frac{\partial \phi_{0,0}^{i}(x,\xi)}{\partial \xi} \\
&= (1-\gamma_i)\xi^{-\gamma_i}F(\alpha_i,\gamma_i,x)F(\alpha_i+1-\gamma_i,2-\gamma_i,\xi) \\
&\quad +\frac{\alpha_i+1-\gamma_i}{2-\gamma_i}\xi^{1-\gamma_i}F(\alpha_i,\gamma_i,x)F(\alpha_i+2-\gamma_i,3-\gamma_i,\xi) \\
&\quad -\frac{\alpha_i}{\gamma_i}x^{1-\gamma_i}F(\alpha_i+1-\gamma_i,2-\gamma_i,x)F(\alpha_i+1,\gamma_i+1,\xi)
\end{aligned}
\tag{5}
$$

$$
\begin{aligned}
\phi_{1,1}^{i}(x,\xi) &= \frac{\partial \phi_{0,1}^{i}(x,\xi)}{\partial x} \\
&= \frac{\alpha_i}{\gamma_i}(1-\gamma_i)\xi^{-\gamma_i}F(\alpha_i+1,\gamma_i+1,x) \\
&\qquad \times F(\alpha_i+1-\gamma_i,2-\gamma_i,\xi) \\
&\quad +\frac{\alpha_i}{\gamma_i}\cdot\frac{\alpha_i+1-\gamma_i}{2-\gamma_i}\xi^{1-\gamma_i}F(\alpha_i+1,\gamma_i+1,x) \\
&\qquad \times F(\alpha_i+2-\gamma_i,3-\gamma_i,\xi) \\
&\quad -\frac{\alpha_i}{\gamma_i}(1-\gamma_i)x^{-\gamma_i}F(\alpha_i+1-\gamma_i,2-\gamma_i,x) \\
&\qquad \times F(\alpha_i+1,\gamma_i+1,\xi) \\
&\quad -\frac{\alpha_i}{\gamma_i}\cdot\frac{\alpha_i+1-\gamma_i}{2-\gamma_i}x^{1-\gamma_i}F(\alpha_i+2-\gamma_i,3-\gamma_i,x) \\
&\qquad \times F(\alpha_i+1,\gamma_i+1,\xi)
\end{aligned}
\tag{6}
$$

Proof: Based on the differential properties of the Kummer function we can know that

$$\frac{d}{dx}F(\alpha,\gamma,x)=\frac{\alpha}{\gamma}F(\alpha+1,\gamma+1,x) \tag{7}$$

So

$$\begin{aligned}
\phi_{1,0}^{i}(x,\xi) &= \frac{\partial \phi_{0,0}^{i}(x,\xi)}{\partial x} \\
&= \xi^{1-\gamma_i}F(\alpha_i+1-\gamma_i,2-\gamma_i,\xi)\frac{\partial}{\partial x}F(\alpha_i,\gamma_i,x) \\
&\quad -(1-\gamma_i)x^{-\gamma_i}F(\alpha_i+1-\gamma_i,2-\gamma_i,x)F(\alpha_i,\gamma_i,\xi) \\
&\quad -x^{1-\gamma_i}F(\alpha_i,\gamma_i,\xi)\frac{\partial}{\partial x}F(\alpha_i+1-\gamma_i,2-\gamma_i,x) \\
&= \frac{\alpha_i}{\gamma_i}\xi^{1-\gamma_i}F(\alpha_i+1,\gamma_i+1,x)F(\alpha_i+1-\gamma_i,2-\gamma_i,\xi) \\
&\quad -(1-\gamma_i)x^{-\gamma_i}F(\alpha_i+1-\gamma_i,2-\gamma_i,x)F(\alpha_i,\gamma_i,\xi) \\
&\quad -x^{1-\gamma_i}\frac{\alpha_i+1-\gamma_i}{2-\gamma_i}F(\alpha_i+2-\gamma_i,3-\gamma_i,x)F(\alpha_i,\gamma_i,\xi)
\end{aligned}$$

Thus: Similarly verifiable (5) and (6).

THE MAIN THEOREM AND ITS PROOF

Theorem If boundary value problem (1) has a unique solution, the left area $(a<x<c)$ is:

$$y_1 = D\cdot\frac{1}{E+\dfrac{1}{R+\varphi(a)}}\cdot\frac{1}{R+\varphi(a)}\cdot\varphi(x) \tag{8}$$

The right area $(c<x<b)$ solution is:

$$y_2 = D \cdot \frac{1}{E + \frac{1}{R + \varphi(a)}} \cdot \frac{1}{R + \varphi(a)} \cdot \frac{\phi_{0,1}^{1}(c,c)}{m\varphi^*(c)\phi_{1,1}^{1}(a,c) - n\phi_{1,0}^{1}(c,b)} \cdot \varphi^*(x) \tag{9}$$

Where $\varphi^*(x)$ is defined as similar kernel function of the right area :

$$\varphi^*(x) = \frac{G\phi_{0,0}^{2}(x,b) + H\phi_{0,1}^{2}(x,b)}{G\phi_{1,0}^{2}(c,b) + H\phi_{1,1}^{2}(c,b)} (c < x < b) \tag{10}$$

Where φ (x) is defined as similar kernel function of the left area :

$$\varphi(x) = \frac{m\varphi^*(c)\ \ \phi_{0,1}^{1}(x,c)\text{-}n\phi_{0,0}^{1}(x,c)}{m\varphi^*(c)\ \ \phi_{1,1}^{1}(a,c)\text{-}n\phi_{1,0}^{1}(a,c)} (a < x < c) \tag{11}$$

Proof: First of all, Based on the lemma 1 we know that the general solution for

$$xy_1'' + (\gamma_1 - x)y_1' - \alpha_1 y_1 = 0(a < x < c) \tag{12}$$

$$xy_2'' + (\gamma_2 - x)y_2' - \alpha_2 y_2 = 0(c < x < b) \tag{13}$$

are as follows

$$y_1 = A_1 F(\alpha_1, \gamma_1, x) + B_1 x^{1-\gamma_1} F(\alpha_1 + 1 - \gamma_1, 2 - \gamma_1, x) \tag{14}$$

$$y_2 = A_2 F(\alpha_2, \gamma_2, x) + B_2 x^{1-\gamma_2} F(\alpha_2 + 1 - \gamma_2, 2 - \gamma_2, x) \tag{15}$$

Where A_1, B_1, A_2, B_2 are real constants.

According the left boundary conditions $[Ey_1 + (1 + ER)y_1']_{x=a} = D$, we can know that

$$\begin{aligned}
&A_1[EF(\alpha_1,\gamma_1,a)+(1+ER)\frac{\alpha_1}{\gamma_1}F(\alpha_1+1,\gamma_1+1,a)] \\
&+B_1[Ea^{1-\gamma_1}F(\alpha_1+1-\gamma_1,2-\gamma_1,a) \\
&+(1+ER)(1-\gamma_1)a^{-\gamma_1}F(\alpha_1+1-\gamma_1,2-\gamma_1,a) \\
&+(1+ER)a^{1-\gamma_1}\frac{\alpha_1+1-\gamma_1}{2-\gamma_1}F(\alpha_1+2-\gamma_1,3-\gamma_1,a)]=D
\end{aligned} \tag{16}$$

According to the continuity conditions $y_1|_{x=c}=my_2|_{x=c}, y'_1|_{x=c}=ny'_2|_{x=c}$ we can get

$$\begin{aligned}
&A_1F(\alpha_1,\gamma_1,c)+B_1c^{1-\gamma_1}F(\alpha_1+1-\gamma_1,2-\gamma_1,c) \\
&-mA_2F(\alpha_2,\gamma_2,c)-mB_2c^{1-\gamma_2}F(\alpha_2+1-\gamma_2,2-\gamma_2,c)=0
\end{aligned} \tag{17}$$

$$\begin{aligned}
&A_1\frac{\alpha_1}{\gamma_1}F(\alpha_1+1,\gamma_1+1,c)-nA_2\frac{\alpha_2}{\gamma_2}F(\alpha_2+1,\gamma_2+1,c) \\
&+B_1[(1-\gamma_1)c^{-\gamma_1}F(\alpha_1+1-\gamma_1,2-\gamma_1,c) \\
&+c^{1-\gamma_1}\frac{\alpha_1+1-\gamma_1}{2-\gamma_1}F(\alpha_1+2-\gamma_1,3-\gamma_1,c)] \\
&-nB_2[(1-\gamma_2)c^{-\gamma_2}F(\alpha_2+1-\gamma_2,2-\gamma_2,c) \\
&+c^{1-\gamma_2}\frac{\alpha_2+1-\gamma_2}{2-\gamma_2}F(\alpha_2+2-\gamma_2,3-\gamma_2,c)]=0
\end{aligned} \tag{18}$$

By the right boundary conditions of the complex confluence hypergeometric equation boundary value problem (1) that $[Gy_2+Hy'_2]_{x=b}=0$ we can get

$$\begin{aligned}
&A_2[GF(\alpha_2,\gamma_2,b)+H\frac{\alpha_2}{\gamma_2}F(\alpha_2+1,\gamma_2+1,b)] \\
&+B_2[Gb^{1-\gamma_2}F(\alpha_2+1-\gamma_2,2-\gamma_2,b) \\
&+H(1-\gamma_2)b^{-\gamma_2}F(\alpha_2+1-\gamma_2,2-\gamma_2,b) \\
&+Hb^{1-\gamma_2}\frac{\alpha_2+1-\gamma_2}{2-\gamma_2}F(\alpha_2+2-\gamma_2,3-\gamma_2,b)]=0
\end{aligned} \tag{19}$$

Getting linear equations about of undetermined coefficients A_1, B_1, A_2 and B_2 by combining equation (16), (17), (18) and (19).

If we assume that complex confluence hypergeometric equation boundary value problem (1) has a unique solution, then there will be a determinant of coefficient of linear equations not be 0,mean that $\Delta \neq 0$.And after calculation using (3), (4), (5), (6) for simplifying we can get

$$\begin{aligned}\Delta = & m[G\phi_{0,0}^{2}(c,b)+H\phi_{0,1}^{2}(c,b)] \\ & \cdot[E\phi_{0,1}^{1}(a,c)+(1+ER)\phi_{1,1}^{1}(a,c)] \\ & -n[G\phi_{1,0}^{2}(c,b)+H\phi_{1,1}^{2}(c,b)] \\ & \cdot[E\phi_{0,0}^{1}(a,c)+(1+ER)\phi_{1,0}^{1}(a,c)]\end{aligned} \tag{20}$$

Using Kramer rule to solve the equations set which is composed of (14), (15), (16) and (17) equations, we can get

$$\begin{aligned}A_1 = & \frac{D}{\Delta}\{m\ [G\ \phi_{0,0}^{2}(c,b)+H\ \phi_{0,1}^{2}(c,b)] \\ & \cdot[(1-\gamma_1)\ c^{-\gamma_1}F(\alpha_1+1-\gamma_1,2-\gamma_1,c) \\ & +c^{1-\gamma_1}\frac{\alpha_1+1-\gamma_1}{2-\gamma_1}F(\alpha_1+2-\gamma_1,3-\gamma_1,c)] \\ & -n\ [G\phi_{1,0}^{2}(c,b)+H\phi_{1,1}^{2}(c,b)] \\ & c^{1-\gamma_1}F(\alpha_1+1-\gamma_1,2-\gamma_1,c)\}\end{aligned} \tag{21}$$

$$\begin{aligned}B_1 = & -\frac{D}{\Delta}\{m[G\phi_{0,0}^{2}(c,b)+H\phi_{0,1}^{2}(c,b)] \\ & \cdot\frac{\alpha_1}{\gamma_1}F(\alpha_1+1,\gamma_1+1,c) \\ & -n[G\phi_{1,0}^{2}(c,b)+H\phi_{1,1}^{2}(c,b)]F(\alpha_1,\gamma_1,c)\}\end{aligned} \tag{22}$$

$$\begin{aligned}A_2 = & \frac{D}{\Delta}[Gb^{1-\gamma_2}F(\alpha_2+1-\gamma_2,2-\gamma_2,b) \\ & +H(1-\gamma_2)b^{-\gamma_2}F(\alpha_2+1-\gamma_2,2-\gamma_2,b) \\ & +Hb^{1-\gamma_2}\frac{\alpha_2+1-\gamma_2}{2-\gamma_2}\end{aligned} \tag{23}$$

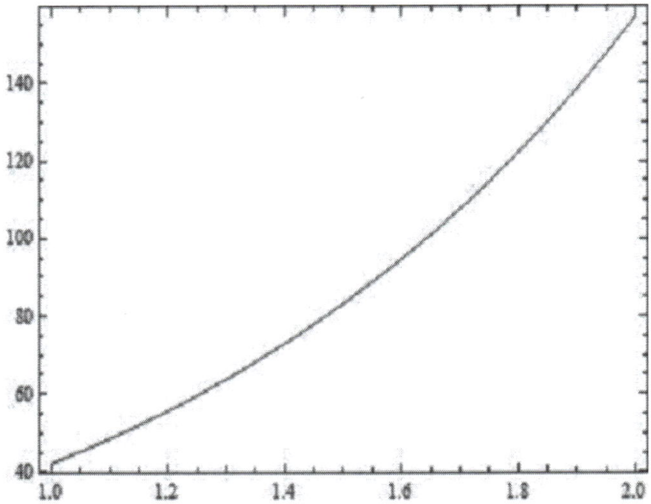

FIGURE 1. Cuver of solution to boundary value problem(25) (left area)

$$\cdot F(\alpha_2+2-\gamma_2,3-\gamma_2,b)]\phi_{0,1}^{1}(c,c)$$

$$B_2=-\frac{D}{\Delta}\begin{bmatrix} GF(\alpha_2,\gamma_2,b) \\ +H\frac{\alpha_2}{\gamma_2}F(\alpha_2+1,\gamma_2+1,b) \end{bmatrix}\phi_{0,1}^{1}(c,c) \tag{24}$$

Then y_1, y_2 can be obtained if we get A_1, B_1, A_2, B_2 substitution of (14), (15). And if we simplify y_1, y_2 using the similar kernel function of the left and the right area, we can get the solution of the left and the right area (8) and (9), respectively.

GIVE AN EXAMPLE

Draw the image of following boundary value problem solution:

$$\begin{cases} xy_1''+(0.5-x)y_1'-y_1=0(1<x<2) \\ xy_2''+(1.5-x)y_2'-y_2=0(2<x<3) \\ [y_1+2y_1']_{x=1}=1 \\ y_1|_{x=2}=y_2|_{x=2} \\ y'_1|_{x=2}=y'_2|_{x=2} \\ [y_2+y'_2]_{x=3}=0 \end{cases} \tag{25}$$

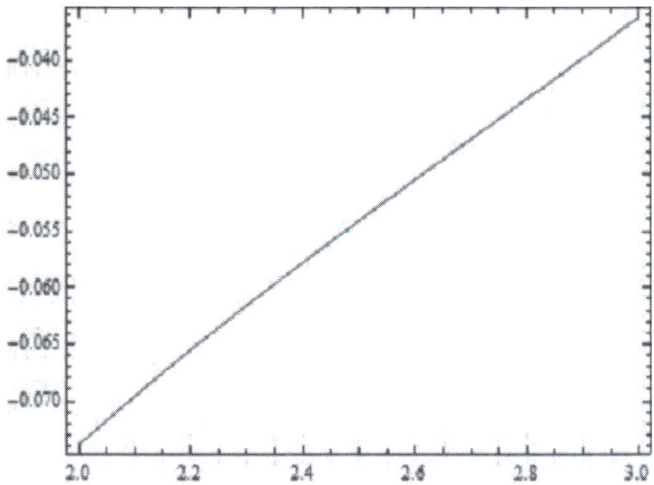

FIGURE 2. Cuver of solution to boundary value problem(25) (right area)

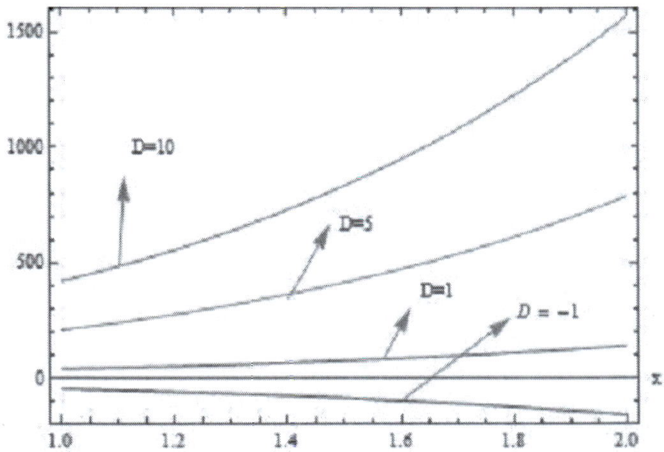

FIGURE 3. The influence of parameter D on curve(right area)

According to the preliminary knowledge we can get the general solution of the target equation as follows.

$$y_1 = A_1 F(1,0.5,x) + B_1 x^{\frac{1}{2}} F(1.5,1.5,x) \tag{26}$$

$$y_2 = A_2 F(1,1.5,x) + B_2 x^{\frac{-1}{2}} F(0.5,0.5,x) \tag{27}$$

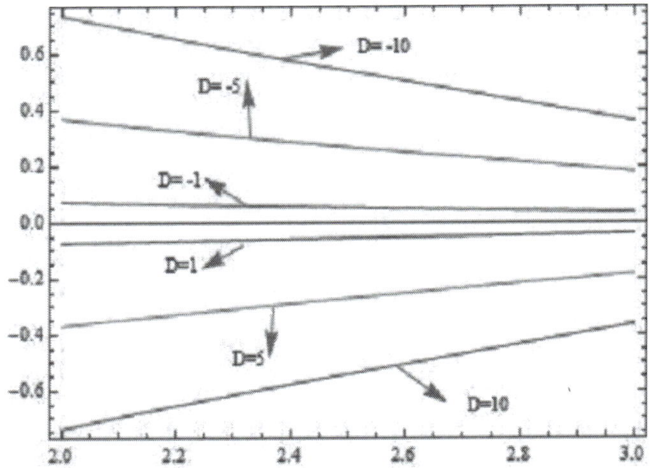

FIGURE 4. The influence of parameter D on curve(right area)

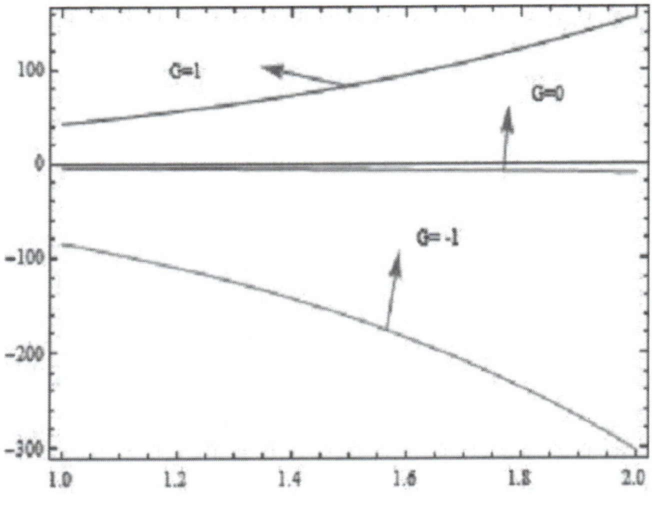

FIGURE 5. The influence of parameter G on curve(left area)

Through the comparison of the toundary value problem (1) and the boundary value problem(25) we found that $\alpha_1 = 1, \alpha_2 = 1, \gamma_1 = 0.5, \gamma_2 = 1.5, D = 1, E = 1, F = 1,$ $, G = 1, H = 1, a = 1, b = 3, c = 2, m = 1, n = 1$ using MATHEMATICA language program to draw the solution of the boundary value problem in the specified image definition domain as follows:

By changing the parameters of D and G in the condition of observation and analysis of the image change.

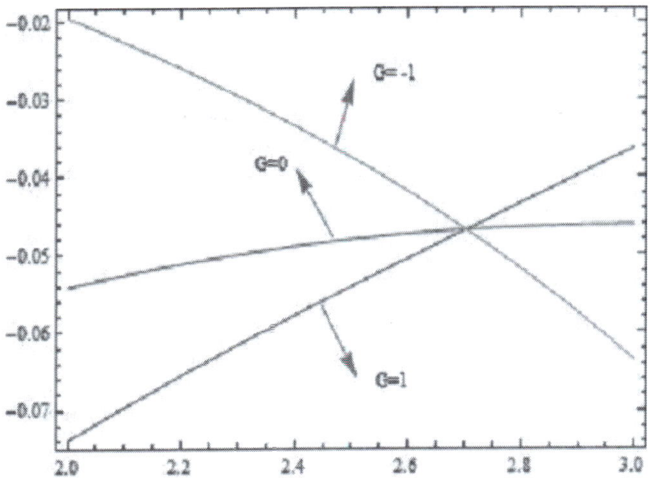

FIGURE 6. The influence of parameter G on curve(right area)

CONCLUSIONS AND UNDERSTANDING

1) With the definition of the left, the right area of similar kernel function (10) and (11), the solutions of boundary value problem (1) can be written as the continued fraction structure. In the progress of solving the boundary value problem for composite Kummer equation, we can get the left area and the right area solution directly by constructing similarity kernel function. This simplifies the solving process, improves the solution efficiency, contributes to the researchers compile corresponding analysis software.
2) From the Figure 3 and Figure 4 we can see that the greater absolute value of the parameter D the greater function curve fluctuations in the left area and the right area.
3) From the Figure 4 we can see that the image function on the X axis symmetry when D were obtained from the two opposite number.
4) From the Figure 5 we can see that the greater absolute value of the parameter G the greater function curve fluctuations in the left area.
5) From the Figure 5 we can see that in the right area regardless of how to change the parameters of the G image through a fixed point and the point coordinates for (2.71,-0.04748).

ACKNOWLEDGMENT

This work is supported the by the Scientific Research Fund of the Sichuan Provincial Education Department of China (Grant No. 12ZA164).

REFERENCES

1. Gaoxiong-Wang, Zhiming-Zhou, Siming-Zhu, Shousong-Wang. Ordinary Differential Equation[M]. Beijing:Higher Education Press, 1978.
2. Li Shun-chu. The Similarity Structuring Method of Boundary Value Problems of the Composite Differential Equations (in Chinese) [J]. Journal of Xihua University (Natural Science Edition), 2013, 32(4): 27-31.
3. Li Shun-chu, Wu Xiao-qing. Several Important Properties for the Boundary Value Problems of Second-order Linear Homogeneous Differential Equation (in Chinese)[J]. Journal of Xihua University (Natural Science Edition), 2013, 32(1): 23-26.
4. LI Shun-Chu, Liao Zhi-Jian. Constructing the Solution of Boundary Value Problem of the Differential Equation with it's an Arbitrary Non-trivial Solution (in Chinese) [J]. Journal of Sichuan University (Natural Science Edition), 2012, 49(6): 1209-1213.
5. WANG Fu-rong, LI Shun-chu, XU Dong-xu. The Similarity Construction Method of a Kind of Boundary-Value Problem of Airy Equation (in Chinese) [J]. Journal of Hubei Normal University (Natural Science), 2013, 33(1):79-85.
6. Bao Xitao, Li Shunchu, Xiao Xuxia, Xu Dongxu. Similar Constructive Method of Solutions of Nonlinear Seepage Model for Homogeneous Reservoir (in Chinese)[J]. NATURAL GAS AND OIL, 2012, 30(65): 47-51.
7. LI Wei, LI Xiaoping, LI Shunchu, LI Quanyong. SIMILAR STRUCTURE OF THE SOLUTIONS OF MATHEMATICAL MODEL FOR THE NONLINEAR FLOW OF FRACTAL COMMINGLED OIL RESERVOIRS (in Chinese) [J]. Petroleum Geology & Oilfield Development in Daqing, 2012, 31(6): 79-83.
8. WANG Jun-chao, LI Shun-chu, XU Li. Spherical Fluids Flow Models in Composite Reservoir Based on Similar Structure of Solution (in Chinese)[J]. Journal of Guilin University of Technology, 2012, 32(4): 624-627.
9. LI Quan-yong, LI Shun-chu, LI Wei, WANG Jun-chao. Study of the Nonlinear Fluid Flow Model in Dual-porosity Media Reservoir Based on Similar Structure[J].CHINESE JOURNAL OF ENGINEERING MATHEMATICS,2013, 30(1):123-130.
10. LI Quan-yong, LI Shun-chu, LI Wei, WANG Jun-chao. Solution to Base on the Similar Structure of the Double Porosity-Multilayer Reservoir(in Chinese) [J]. College Mathematics, 2013, 29(1): 56-60.
11. LI Shun-chu, Huang Bing-guang. Laplace transform and Bessel functions and the theoretical basis of well test analysis [M].Beijing: Petroleum Industry Press, 2000.

12. LI Shun-chu, WANG Jun-chao, XU Li. The Similar Structure of Solutions to the Spherical Fluids Flow Problems for a Composite Reservoir (in Chinese) [J]. Journal of Mathematics in Practice and Theory, 2014,44(3):122-127.
13. Li Ke, Li Shunchu, Dong Yanan. The Solution of Well Test Analysis Model Based on Similar Structure in the Multilayer Reservoir (in Chinese)[J]. Well Testing, 2013, 22(2):7-9.
14. Xiao Xuxia, Li Shunchu. The Similar Structure of Solutions for the Boundary Value Problem of Euler's Hypergeometric Differential Equation (in Chinese) [J]. Journal of Inner Mongolia Normal University (Natural Science Edition), 2012, 41(6): 597-600(-603).
15. BAO Xi-tao, LI Shun-chu, LIAO Zhi-Jian. Similar Constructive Method of Solution for the Boundary Value Problem of Euler Hyper-geometric Differential Equation (in Chinese)[J]. Journal of Southwest University of Science and Technology, 2012, 27(4):101-105.

6

ANALYTIC SOLUTION OF LINEAR FRACTIONAL DIFFERENTIAL EQUATION WITH JUMARIE DERIVATIVE IN TERM OF MITTAG-LEFFLER FUNCTION

Uttam Ghosh[1*], Srijan Sengupta[2], Susmita Sarkar[2], Shantanu Das[3,4]

[1] Department of Mathematics, Nabadwip Vidyasagar College, Nabadwip, Nadia, West Bengal, India

[2] Department of Applied Mathematics, University of Calcutta, Kolkata, West Bengal, India

[3] Reactor Control System Design Section Bhabha Atomic Research Centre, Mumbai, India

[4] Department of Physics, Jadavpur University Kolkata, West Bengal, India

ABSTRACT

There is no unified method to solve the fractional differential equation. The type of derivative here used in this paper is of Jumarie formulation, for the several differential equations studied. Here we develop an algorithm to solve the linear fractional differential equation composed via Jumarie fractional derivative in terms of Mittag-Leffler function; and show its conjugation with ordinary calculus. In these fractional differential equations the one parameter Mittag-Leffler function plays the role similar as exponential function used in ordinary differential equations.

INTRODUCTION

The analytical solutions of the fractional differential equation are emerging branch of applied science also in basic science. Different methods are developing to solve the fractional differential equations. Since the definition of fractional derivative is modifying to relate it with the classical derivative. Mathematicians are trying to develop the formulas of fractional calculus but geometry of fractional derivative has no concrete shape[1]. Depending on different type of derivatives different methods of solution are developing [2-7]. Riemann-Liouville definition the fractional derivative of a constant is non-zero which creates a difficulty to relate between the classical calculus. To overcome this difficulty Jumarie [2,3,4,5] modified the definition of fractional derivative of Riemann-Liouvell type and with this new formulation, we obtain the derivative of a constant as zero. Thus using this definition linking between the fractional and classical calculus becomes easier. There is no unique method to solve the linear fractional differential equations. Using the Jumarie modified definition of fractional derivative we obtain the derivative of Mittag-Leffler function as Mittal-Leffler function, as in case of classical whole-order derivative the derivative of $\exp(x)$ is itself exponential function. Thus via use of Jumarie modified Riemann-Liouvelli derivative, there exists conjugation with classical calculus, which eases in many cases to solve fractional differential equation composed with Jumarie fractional derivative. Here we want to develop an algorithm to solving the linear fractional differential equation using the Mittag-Leffler function. We have obtained applied this method to homogeneous fractional differential equations and got corresponding fundamental solution.

Organization of the paper is as follows. In section 2.0 some definition of fractional derivative is reproduced with essential examples. In section 3.0 and 4.0 some properties of Mittag-Leffler function is described. Finally in section 5.0 the methods for solving the linear fractional differential equation composed by Jumarie fractional derivative is developed using the Mittag-Leffler function.

Some Definitions of Fractional

There are many definition of fractional derivative. Grunwald-Letnikov fractional derivative [6] Liouville fractional derivative [8], Riemann-Liouville fractional derivative [10] Caputo fractional derivative [8,9,10,11], Kolwanker-Gangal local fractional derivative [12,13,14,15,16], Jumarie modified fractional derivative [2], Here we use the Riemann-Liouville fractional derivative and its modified form by Jumarie [2],

Riemann-Liouville Definition of Fractional Derivative

Let the function $f(t)$ is one time integrable then the integro-differential expression

$$_aD_t^{\alpha} f(t) = \frac{1}{\Gamma(-\alpha+m+1)}\left(\frac{d}{dt}\right)^{m+1} \int_a^t (t-\tau)^{m-\alpha} f(\tau)d\tau$$

is known as the Riemann-Liouville (R-L)definition of fractional derivative [6] with m as integer with condition $m \le \alpha < m+1$.

In Riemann-Liouville definition the function, $f(t)$ is integrated $(m-\alpha)$ fold and then differentiate $m+1$ times. We can re-write the above as follows

The left R-L fractional derivative is defined by

$$ {}_aD_t^{\alpha} f(t) = \frac{1}{\Gamma(k-\alpha)} \left(\frac{d}{dt} \right)^k \int_a^t (t-\tau)^{k-\alpha-1} f(\tau)d\tau. $$

And the right R-L derivative is

$$ {}_tD_b^{\alpha} f(t) = \frac{1}{\Gamma(k-\alpha)} \left(-\frac{d}{dt} \right)^k \int_t^b (\tau-t)^{k-\alpha-1} f(\tau)d\tau $$

Where in above k is integer such that $(k-1) \le \alpha < k$ that is just greater than fractional number α.

Using the left R-L derivative we get the fractional derivative of the function $f(t) = K$ as non-zero, as demonstrated below.

$$ {}_aD_t^{\alpha} f(t) = \frac{1}{\Gamma(1-\alpha)} \frac{d}{dt} \int_a^t (t-\xi)^{-\alpha} K d\xi $$

$$ = -\frac{K}{\Gamma(1-\alpha)} \frac{d}{dt} \left. \frac{(t-\xi)^{1-\alpha}}{1-\alpha} \right]_a^t $$

$$ = K \frac{(t-a)^{1-\alpha}}{\Gamma(1-\alpha)} $$

Similarly the right R-L derivative of $f(t) = K$ is

$$ {}_tD_b^{\alpha} f(t) = K \frac{(b-t)^{1-\alpha}}{\Gamma(1-\alpha)}. $$

This shows that the fractional derivative of a constant (K) is non-zero but in classical calculus derivative of a constant is zero which is contradiction between the classical derivative and the fractional derivative of a constant. To overcome this difference Jumarie [2] modified the left R-L fractional derivative.

We get the R-L left derivative of a power function as

Jumarie Modified Definition of the Fractional Derivative Is

$$D_t^{\alpha} f(t) \overset{\text{def}}{=} \begin{cases} \dfrac{1}{\Gamma(-\alpha)} \displaystyle\int_0^t (t-\xi)^{-\alpha-1} f(\xi) d\xi, \ \alpha < 0. \\ \dfrac{1}{\Gamma(1-\alpha)} \dfrac{d}{dt} \displaystyle\int_0^t (t-\xi)^{-\alpha} \begin{bmatrix} f(\xi) \\ -f(0) \end{bmatrix} d\xi, \ 0 < \alpha < 1. \\ [f^{(\alpha-n)}(t)]^{(n)}, \quad n \le \alpha < n+1, n \ge 1. \end{cases}$$

Using this definition we get $D^{\alpha}\{K\} = 0, \ 0 \le \alpha < 1$. The above formula in line-1, becomes fractional order integration if we replace α by $-\alpha$ which is

$${}_aD_t^{-\alpha} f(t) = \frac{1}{\Gamma(\alpha)} \int_a^t (t-\tau)^{\alpha-1} f(\tau) d\tau$$

Using the above formula we get for $f(t) = (t-a)^{\gamma}$, the fractional integral for order α

$${}_aD_t^{-\alpha} (t-a)^{\gamma} = \frac{1}{\Gamma(\alpha)} \int_a^t (t-\tau)^{\alpha-1} (\tau-a)^{\gamma} d\tau$$

Using the substitution $\tau = a + \xi(t-a)$ we have for; $\tau = a$, $\xi = 0$ and for $\tau = t$, $\xi = 1$; $d\tau = (t-a)d\xi$, $(t-\tau) = t - a - \xi(t-a) = (t-a)(1-\xi)$; $(\tau - a) = \xi(t-a)$, we get the following

$$
\begin{aligned}
{}_aD_t^{-\alpha}(t-a)^{\gamma} &= \frac{1}{\Gamma(\alpha)}\int_a^t (t-\tau)^{\alpha-1}(\tau-a)^{\gamma}\,d\tau \\
&= \frac{1}{\Gamma(\alpha)}\int_0^1 (t-a)^{\alpha-1}(1-\xi)^{\alpha-1}\xi^{\gamma}(t-a)^{\gamma}(t-a)d\xi \\
&= \frac{(t-a)^{\gamma+\alpha}}{\Gamma(\alpha)}\int_0^1 \xi^{\gamma}(1-\xi)^{\alpha-1}d\xi \\
&= \frac{(t-a)^{\gamma+\alpha}}{\Gamma(\alpha)}\mathrm{B}(\alpha,\gamma+1) \\
&= \frac{\Gamma(\gamma+1)}{\Gamma(\gamma+1+\alpha)}(t-a)^{\gamma+\alpha}, \quad (\alpha<0,\gamma>-1)
\end{aligned}
$$

We used Beta-function $= \dfrac{\Gamma(\alpha)\Gamma(\gamma+1)}{\Gamma(\alpha+\gamma+1)}$ defined as

$$\mathrm{B}(\alpha,\gamma+1) = \int_0^1 \xi^{\gamma}(1-\xi)^{\alpha-1}d\xi$$

$$\mathrm{B}(p,q) \overset{\text{def}}{=} \int_0^1 u^{p-1}(1-u)^{q-1}\,du = \frac{\Gamma(p)\Gamma(q)}{\Gamma(p+q)}.$$

Applying the above obtained result the fractional integral of order $(1-\upsilon)$, with $0 \le \upsilon < 1$ is

$${}_aD_t^{-(1-\upsilon)}(t-a)^{\gamma} = \frac{\Gamma(\gamma+1)}{\Gamma(\gamma+2-\upsilon)}(t-a)^{\gamma+1-\upsilon}.$$

Taking one whole derivative of the above we get

$$D^1\left[{}_aD_t^{-(1-\upsilon)}(t-a)^\gamma\right] = {}_aD_t^{\upsilon}(t-a)^\gamma$$

$$= \frac{d}{dt}\left[\frac{\Gamma(\gamma+1)}{\Gamma(\gamma+2-\upsilon)}(t-a)^{\gamma+1-\upsilon}\right]$$

$$= \frac{\Gamma(\gamma+1)}{\Gamma(\gamma-\upsilon+2)}(\gamma+1-\upsilon)(t-a)^{\gamma-\upsilon}$$

$$= \frac{\Gamma(\gamma+1)}{(\gamma-\upsilon+1)\Gamma(\gamma+1-\upsilon)}(\gamma+1-\upsilon)(t-a)^{\gamma-\upsilon}$$

$$= \frac{\Gamma(\gamma+1)}{\Gamma(\gamma+1-\upsilon)}(t-a)^{\gamma-\upsilon}$$

We have therefore calculated fractional derivative by R-L left formula for υ such that $0 \le \upsilon < 1$ thus our nearest integer is one that is $k = 1$ and we write that below

$$_aD_t^{\upsilon}f(t) = \frac{1}{\Gamma(1-\upsilon)}\left(\frac{d}{dt}\right)\int_a^t (t-\tau)^{-\upsilon} f(\tau)d\tau$$

$$= D^1\left[D_t^{-(1-\upsilon)}f(t)\right].$$

Thus for $a = 0$, the fractional RL derivative of $f(t) = t^\gamma$ is

$$_0D_t^{\upsilon}t^\gamma = \frac{\Gamma(\gamma+1)}{\Gamma(\gamma+1-\upsilon)}t^{\gamma-\upsilon}$$

For a constant function $f(t) = 1$, putting in above expression $\gamma = 0$, we get

$$_0D_t^{\upsilon}[1] = \frac{1}{\Gamma(1-\upsilon)}t^{-\upsilon}$$

Let us now see what Jumarrie derivative is from above R-L derivative obtained for $f(t) = t^\gamma$. The composition of the Jumarie derivative, with start point of integration as $t = a$ and $f(a) = a^\gamma$ is

$$f^{(\alpha)}[t^{\gamma}]_a^t = \frac{1}{\Gamma(1-\alpha)}\frac{d}{dt}\int_a^t (t-\xi)^{-\alpha}\left[\xi^{\gamma} - a^{\gamma}\right]d\xi$$

$$D^1\left[{}_aD_t^{-(1-\alpha)}t^{\gamma} - {}_aD_t^{-(1-\alpha)}a^{\gamma}\right]$$

$$= {}_aD_t^{\alpha}t^{\gamma} - {}_aD_t^{\alpha}a^{\gamma}.$$

The above expression show that this Jumarrie derivative is composed of two RL derivatives those are ${}_aD_t^{\alpha}t^{\gamma}$ minus RL derivative of a constant ${}_aD_t^{\alpha}a^{\gamma}$. Going by similar steps as done for ${}_0D_t^{\alpha}t^{\gamma}$, we get first the fractional integral in terms of incomplete Gamma function as

$${}_aD_t^{-(1-\alpha)}t^{\gamma} = \frac{t^{\gamma+1-\alpha}}{\Gamma(1-\alpha)}\int_0^{(t-a)/t} z^{-\alpha}(1-z)^{\gamma}\,\mathrm{d}z$$

$$= \frac{t^{\gamma+1-\alpha}}{\Gamma(1-\alpha)}\mathrm{B}_{\eta}(1-\alpha,\gamma+1) \quad \eta = \frac{t-a}{t}.$$

The fractional derivative of t^{γ} is by taking one whole derivative of above expression we get the following

$${}_aD_t^{\alpha}t^{\gamma} = \frac{d}{dt}\left[\frac{t^{\gamma+1-\alpha}}{\Gamma(1-\alpha)}\mathrm{B}_{\eta}(1-\alpha,\gamma+1)\right] \eta = \frac{t-a}{t}.$$

The fractional derivative of a^{γ} is

$${}_aD_t^{\alpha}a^{\gamma} = \frac{a^{\gamma}}{\Gamma(1-\alpha)}(t-a)^{-\alpha}.$$

Therefore

$$f^{(\alpha)}[t^{\gamma}]_a^t = \frac{1}{\Gamma(1-\alpha)}\frac{d}{dt}\int_a^t (t-\xi)^{-\alpha}\left[\xi^{\gamma} - a^{\gamma}\right]d\xi$$

$$= D^1\left[{}_aD_t^{-(1-\alpha)}t^{\gamma} - {}_aD_t^{-(1-\alpha)}a^{\gamma}\right] = {}_aD_t^{\alpha}t^{\gamma} - {}_aD_t^{\alpha}a^{\gamma}$$

$$= \frac{d}{dt}\left[\frac{t^{\gamma+1-\alpha}}{\Gamma(1-\alpha)} \mathrm{B}_{\eta}(1-\alpha, \gamma+1)\right] - \frac{a^{\gamma}}{\Gamma(1-\alpha)}(t-a)^{-\alpha}$$

For $a = 0$, we have

$$f^{(\alpha)}[t^{\gamma}]_0^t = \frac{1}{\Gamma(1-\alpha)} \frac{d}{dt} \int_0^t (t-\xi)^{-\alpha}\left[\xi^{\gamma} - 0\right] d\xi$$

$$= D^1\left[{}_0D_t^{-(1-\alpha)} t^{\gamma}\right] = {}_0D_t^{\alpha} t^{\gamma} = \frac{\Gamma(\gamma+1)}{\Gamma(\gamma+1-\alpha)}(t)^{\gamma-\alpha}$$

We will be using Jumarie derivative for power function t^{γ} with start point of differentiation as $a = 0$, in subsequent sections. When start point of differentiation is non-zero we will be shifting the origin to that non-zero point and use the above formula.

MITTAG-LEFFLER FUNCTION

In 1903 Mittag-Leffler [17,18,19,20] was introduce a function defined by $E_{\alpha}(z) = \sum_{k=0}^{\infty} \frac{z^k}{\Gamma(1+\alpha k)}$, $(z \in \mathbb{C}, \mathrm{Re}(\alpha) > 0)$ is named as the one parameter Mittag-Leffler function. For $\alpha = 1 \text{ and } 2$ we get

$$E_1(z) \overset{\text{def}}{=} 1 + \frac{z}{\Gamma(1+1)} + \frac{z^2}{\Gamma(1+2)} + \frac{z^3}{\Gamma(1+3)} + \ldots\ldots\ldots\ldots\infty$$

$$= 1 + \frac{z}{1!} + \frac{z^2}{2!} + \frac{z^3}{3!} + \ldots\ldots\ldots\ldots\infty = e^z$$

$$E_2(z) \overset{\text{def}}{=} 1 + \frac{z}{\Gamma(1+2)} + \frac{z^2}{\Gamma(1+4)} + \frac{z^3}{\Gamma(1+6)} + \ldots\ldots\ldots\ldots\infty$$

$$= 1 + \frac{z}{2!} + \frac{z^2}{4!} + \frac{z^3}{6!} + \ldots\ldots\ldots\ldots\infty$$

$$= Cosh(\sqrt{z})$$

The integral representation of the Mittag-Leffler function is $E_\alpha(z)=\frac{1}{2\pi}\int_C \frac{t^{\alpha-1}e^t}{t^\alpha - z}dt, (z\in\mathbb{C}, \mathrm{Re}(\alpha)>0)$. Here the path of the integral C is a loop which starts and ends at $-\infty$ and encloses the circles of disk $|t|\leq|z|^{1/\alpha}$ in positive sense : $|\arg(t)|\leq\pi$ on C.

The two parameter Mittag-Leffler function [21] was defined by

$$E_{\alpha,\beta}(z)=\sum_{k=0}^{\infty}\frac{z^k}{\Gamma(\beta+\alpha k)}, (z,\beta\in\mathbb{C}, \mathrm{Re}(\alpha)>0).$$

$$\text{Here}, \mathrm{E}_{\alpha,1}(z)=\mathrm{E}_\alpha(z)=e^z (z\in\mathbb{C}, \mathrm{Re}(\alpha)>0)$$

$$\mathrm{E}_{1,2}(z)=\frac{e^z-1}{z}, \mathrm{E}_{2,2}(z)=\frac{Sinh(\sqrt{z})}{\sqrt{z}}.$$

and the corresponding integral representation of the two parameter Mittag-Leffler function is $E_{\alpha,\beta}(z)=\frac{1}{2\pi}\int_C \frac{t^{\alpha-\beta}e^t}{t^\alpha - z}dt,\ (z\in\mathbb{C}, \mathrm{Re}(\alpha)>0)$. where the contour C is already defined.

Some Properties Mittag-Leffler Function and Its Application

We rewrite the Mittag-Leffler [17, 18, 19] function in the following form by an infinite series

$$E_\alpha(at^\alpha) \overset{\text{def}}{=} 1+\frac{at^\alpha}{\Gamma(1+\alpha)}+\frac{a^2t^{2\alpha}}{\Gamma(1+2\alpha)} +\frac{a^3t^{3\alpha}}{\Gamma(1+3\alpha)}+\ldots\ldots\ldots\ldots\ldots$$

is the one parameter Mittag-Leffler function.

Using Jumarie derivative of order α, with $0\leq\alpha<1$ with start point as $a=0$ for $f(t)=t^{n\alpha}$, that is $(t)^{\alpha(n-1)}\Gamma(n\alpha+1)/\Gamma[\alpha(n-1)+1]$, for $n=1,2,3,\ldots$; and also using Jumarie derivative of constant as zero, we get the following very useful identity. In all the subsequent sections we will say D^α is the Jumarie derivative with zero as start point

$$D^{\alpha}(E_{\alpha}(at^{\alpha}))$$

$$= D^{\alpha}\begin{pmatrix} 1+\dfrac{at^{\alpha}}{\Gamma(1+\alpha)}+\dfrac{a^2t^{2\alpha}}{\Gamma(1+2\alpha)} \\ +\dfrac{a^3t^{3\alpha}}{\Gamma(1+3\alpha)}+\ldots\ldots\ldots\ldots \end{pmatrix}$$

$$= 0+\frac{\Gamma(1+\alpha)a}{\Gamma(1)\Gamma(1+\alpha)}+\frac{\Gamma(1+2\alpha)a^2t^{\alpha}}{\Gamma(1+2\alpha)\Gamma(1+\alpha)}$$

$$+\frac{\Gamma(1+3\alpha)a^3t^{2\alpha}}{\Gamma(1+3\alpha)\Gamma(1+2\alpha)}+\ldots\ldots\ldots$$

$$= a\begin{pmatrix} 1+\dfrac{at^{\alpha}}{\Gamma(1+\alpha)}+\dfrac{a^2t^{2\alpha}}{\Gamma(1+2\alpha)} \\ +\dfrac{a^3t^{3\alpha}}{\Gamma(1+3\alpha)}+\ldots\ldots\ldots\ldots \end{pmatrix}$$

$$= aE_{\alpha}(at^{\alpha})$$

Thus

$$D^{\alpha}(E_{\alpha}(at^{\alpha})) = aE_{\alpha}(at^{\alpha}) \tag{1}$$

This shows that $AE_{\alpha}(at^{\alpha})$ is a solution is a solution of the fractional differential equation

$$D^{\alpha}y = ay \tag{2}$$

Where A is arbitrary constant.
Therefore

$$D^{\alpha}y = ay$$

with $y(0) = 1$ has solution

$$y = E_{\alpha}(at^{\alpha}).$$

Using this property of the Mittag-Leffler one can easily prove the following theorem.

Again integrating the Mittag-Leffler function in the interval [0, x] we get

$$
{}_0D_a^{-\alpha}(E_\alpha(at^\alpha)) = D^{-\alpha}\left(1+\frac{at^\alpha}{\Gamma(1+\alpha)}+\frac{a^2t^{2\alpha}}{\Gamma(1+2\alpha)}+\frac{a^3t^{3\alpha}}{\Gamma(1+3\alpha)}+\cdots\cdots\right)
$$

$$
= \frac{t^\alpha}{\Gamma(1+\alpha)}+\frac{\Gamma(1+\alpha)at^{2\alpha}}{\Gamma(1+2\alpha)\Gamma(1+\alpha)}+\frac{\Gamma(1+2\alpha)a^2t^{3\alpha}}{\Gamma(1+3\alpha)\Gamma(1+2\alpha)}+\cdots\cdots
$$

$$
= \frac{1}{a}\left(1+\frac{at^\alpha}{\Gamma(1+\alpha)}+\frac{a^2t^{2\alpha}}{\Gamma(1+2\alpha)}+\frac{a^3t^{3\alpha}}{\Gamma(1+3\alpha)}+\cdots\cdots\infty-1\right)
$$

$$
= \frac{1}{a}\{E_\alpha(at^\alpha)-1\}
$$

Theorem 1: The Mittag-Leffler function $E_\alpha(at^\alpha)$ satisfies the relation

$$
E_\alpha(at^\alpha)E_\alpha(bt^\alpha) = E_\alpha((a+b)t^\alpha)
$$

Proof: Let $y = E_\alpha(at^\alpha)E_\alpha(bt^\alpha)$ then

$E_\alpha(at^\alpha)E_\alpha(bt^\alpha) = E_\alpha((a+b)t^\alpha)$

Proof: Let $y = E_\alpha(at^\alpha)E_\alpha(bt^\alpha)$ then

$$
\begin{aligned}
y &= E_\alpha(at^\alpha)E_\alpha(bt^\alpha) \\
D^\alpha y &= E_\alpha(bt^\alpha)D^\alpha(at^\alpha)+E_\alpha(at^\alpha)D^\alpha(bt^\alpha) \\
D^\alpha y &= aE_\alpha(at^\alpha)E_\alpha(bt^\alpha)+bE_\alpha(at^\alpha)E_\alpha(bt^\alpha) \\
&= (a+b)E_\alpha(at^\alpha)E_\alpha(bt^\alpha) \\
D^\alpha y &= (a+b)y
\end{aligned} \tag{3}
$$

Using the solution of the equation (2) we get the solution of the equation (3) in the following form

$$
y = AE_\alpha\left([a+b]t^\alpha\right).
$$

From the definition of y we get $y(0)=1$. Therefore we have $y = E_\alpha((a+b)t^\alpha)$.

Thus we get

$$E_\alpha((a+b)t^\alpha) = E_\alpha(at^\alpha)E_\alpha(bt^\alpha) \tag{4}$$

We get useful property of one parameter Mittag-Leffler function.
Using the above property of Mittag-Leffler function we get

$$E_\alpha(at^\alpha)E_\alpha(-at^\alpha) = 1 \text{ or } E_\alpha(-at^\alpha) = \frac{1}{E_\alpha(at^\alpha)}. \tag{5}$$

$$E_\alpha(at^\alpha)E_\alpha(at^\alpha) = E_\alpha(2at^\alpha) \tag{6}$$

COMPLEX MITTAG-LEFFLER FUNCTION AND ITS PROPERTIES

Jumarie [22] defined the complex Mittag-Leffler in the following form

$$E_\alpha(it^\alpha) \overset{\text{def}}{=} \cos_\alpha(t^\alpha) + i\sin_\alpha(t^\alpha)$$

$$\cos_\alpha(t^\alpha) = \frac{E_\alpha(it^\alpha) + E_\alpha(-it^\alpha)}{2} = \sum_{k=1}^{\infty}(-1)^k \frac{t^{2k\alpha}}{(2k\alpha)!}$$

$$\sin_\alpha(t^\alpha) = \frac{E_\alpha(it^\alpha) - E_\alpha(-it^\alpha)}{2} = \sum_{k=1}^{\infty}(-1)^k \frac{t^{(2k+1)\alpha}}{(2k\alpha+\alpha)!}$$

On the other hand Jumarie [22] defined period (M_α) of the function $E_\alpha(it^\alpha)$ in the following form, taking $E_\alpha(i(M_\alpha)^\alpha) = 1$ and therefore

$$\cos_\alpha(t+M_\alpha)^\alpha = \cos_\alpha(t^\alpha), \sin_\alpha(t+M_\alpha)^\alpha = \sin_\alpha(t^\alpha)$$

$$\cos_\alpha((-t)^\alpha) = \cos_\alpha(t^\alpha), \sin_\alpha((-t)^\alpha) = (-1)^\alpha \sin_\alpha(t^\alpha).$$

The series presentation of $\cos_\alpha(t^\alpha)$ is

$$\cos_\alpha(t^\alpha) = 1 - \frac{t^{2\alpha}}{\Gamma(1+2\alpha)} + \frac{t^{4\alpha}}{\Gamma(1+4\alpha)} - \frac{t^{6\alpha}}{\Gamma(1+6\alpha)} + \dots..$$

Taking term by term Jumarie derivative we get

$$D^{\alpha}[\cos_{\alpha}(t^{\alpha})]$$
$$= 0 - \frac{\Gamma(1+2\alpha)t^{2\alpha-\alpha}}{\Gamma(1+2\alpha)\Gamma(1+\alpha)} + \frac{\Gamma(1+4\alpha)t^{4\alpha-\alpha}}{\Gamma(1+4\alpha)\Gamma(1+3\alpha)}$$
$$- \frac{\Gamma(1+6\alpha)t^{6\alpha-\alpha}}{\Gamma(1+6\alpha)\Gamma(1+5\alpha)} + ...$$
$$= -\left[\frac{t^{\alpha}}{\Gamma(1+\alpha)} - \frac{t^{3\alpha}}{\Gamma(1+3\alpha)} + ...\right]$$
$$= -\sin_{\alpha}(t^{\alpha})$$

SOLUTION OF LINEAR SECOND ORDER FRACTIONAL DIFFERENTIAL EQUATION

Let us consider the function

$$y = AE_{\alpha}(at^{\alpha}) + BE_{\alpha}(bt^{\alpha})$$

with A and B is constants. Differentiating $\alpha-$ times with respect to t, for $0 < \alpha < 1$, with Jumarie derivative we get

$$D^{\alpha}y = AaE_{\alpha}(at^{\alpha}) + BbE_{\alpha}(bt^{\alpha})$$
$$D^{\alpha}y - ay = AaE_{\alpha}(at^{\alpha}) + BbE_{\alpha}(bt^{\alpha}) - ay$$
$$= AaE_{\alpha}(at^{\alpha}) + BbE_{\alpha}(bt^{\alpha})$$
$$- a\left(AE_{\alpha}(at^{\alpha}) + BE_{\alpha}(bt^{\alpha})\right)$$
$$= B(b-a)E_{\alpha}(bt^{\alpha})$$
$$D^{\alpha}y - ay = B(b-a)E_{\alpha}(bt^{\alpha})$$

Differentiating above by Jumarrie derivative and re-arranging, we get

$$D^{2\alpha}y - aD^{\alpha}y = Bb(b-a)E_{\alpha}(bt^{\alpha})$$
$$D^{2\alpha}y - (a+b)D^{\alpha}y + aby = 0$$

This shows that the fractional differential equation

$$D^{2\alpha}y - (a+b)D^{\alpha}y + aby = 0$$

has solution in the form

$$y = AE_{\alpha}(at^{\alpha}) + BE_{\alpha}(bt^{\alpha}).$$

On the other hand consider the differential equation

$$D^{2\alpha}y - (a+b)D^{\alpha}y + aby = 0$$

it can be express in the following form

$$(D^{\alpha} - a)(D^{\alpha} - b)y(t) = 0. \tag{7}$$

Let, $(D^{\alpha} - b)y(t) = x(t)$ then equation (7) reduce to the form

$$(D^{\alpha} - a)x(t) = 0 \quad \text{or} \quad D^{\alpha}x(t) = ax(t)$$

Solution of the above equation is same as the solution of the equation (2) which is

$$x(t) = A_1E_{\alpha}(at^{\alpha})$$
$$(D^{\alpha} - b)y(t) = A_1E_{\alpha}(at^{\alpha})$$
$$D^{\alpha}y - by = A_1E_{\alpha}(at^{\alpha})$$
$$E_{\alpha}(-bt^{\alpha})(D^{\alpha}y - by) = A_1E_{\alpha}(at^{\alpha})E_{\alpha}(-bt^{\alpha})$$
$$D^{\alpha}(yE_{\alpha}(-bt^{\alpha})) = \frac{A_1}{a-b}D^{\alpha}(E_{\alpha}(at^{\alpha})E_{\alpha}(-bt^{\alpha}))$$

On integrating both side we get that is applying $D^{-\alpha}$ on both sides of above, we get

$$yE_{\alpha}(-bt^{\alpha}) = \frac{A_1}{a-b}(E_{\alpha}(at^{\alpha})E_{\alpha}(-bt^{\alpha})) + B$$
$$y = AE_{\alpha}(at^{\alpha}) + BE_{\alpha}(bt^{\alpha}), \quad \text{where} \quad A = \frac{A_1}{a-b}.$$

Therefore $y = AE_{\alpha}(at^{\alpha}) + BE_{\alpha}(bt^{\alpha})$ is a solution of the differential equation. Thus we can state the following theorem

Theorem 2: The fractional differential equation

$$(D^{\alpha} - a)(D^{\alpha} - b)y(t) = 0$$

has solution of the form

$$y = AE_{\alpha}(at^{\alpha}) + BE_{\alpha}(bt^{\alpha})$$

where A and B are constants.

Proof of the theorem is follows from the previous arguments.

Similarly one can generalized the solution of the differential equation in the following form

If

$$(D^{\alpha} - a_1)(D^{\alpha} - a_2)(D^{\alpha} - a_3)\ldots(D^{\alpha} - a_n)y(t) = 0$$

with all a_i's are distinct be a fractional differential equation with $0 \le \alpha < 1$ then solution of the differential equation will be

$$y = \sum_{i=1}^{n} A_i E_{\alpha}(a_i t^{\alpha})$$

where A_i are arbitrary constants and $E_{\alpha}(a_i t^{\alpha})$ is one parameter Mittag-Leffler function.

Let us consider the function

$$y = (At^{\alpha} + B)E_{\alpha}(at^{\alpha})$$

Where A and B are constants.

Then

$$D^{\alpha}y = \Gamma(1+\alpha)AE_{\alpha}(at^{\alpha}) + (At^{\alpha} + B)aE_{\alpha}(at^{\alpha})$$

$$\begin{aligned} D^{\alpha}y - ay &= \Gamma(1+\alpha)AE_{\alpha}(at^{\alpha}) \\ &\quad + (At^{\alpha} + B)aE_{\alpha}(at^{\alpha}) - a\{(At^{\alpha} + B)E_{\alpha}(at^{\alpha})\} \\ &= \Gamma(1+\alpha)AE_{\alpha}(at^{\alpha}) \end{aligned}$$

$$\begin{aligned} D^{2\alpha}y - aD^{\alpha}y &= a\Gamma(1+\alpha)AE_{\alpha}(at^{\alpha}) \\ &= a(D^{\alpha}y - ay) \end{aligned}$$

$$D^{2\alpha}y - 2aD^{\alpha}y + a^2y = 0$$

Thus solution of the differential equation

$$D^{2\alpha}y - 2aD^{\alpha}y + a^2y = 0$$

is

$$y = (At^{\alpha} + B)E_{\alpha}(at^{\alpha})$$

A and *B* are constants.

On the other hand consider the differential equation

$$(D^{\alpha} - a)^2 y = 0 \quad \text{or} \quad (D^{2\alpha} - 2aD^{\alpha} + a^2)y = 0$$
$$D^{2\alpha}y - 2aD^{\alpha}y + a^2y = 0 \tag{8}$$

Let $(D^{\alpha} - a)y = v$ then equation (8) reduce to the form

$$(D^{\alpha} - a)v = 0 \tag{9}$$

Solution of this differential equation is $v(t) = A_1E_{\alpha}(at^{\alpha})$

$$(D^{\alpha} - a)y = A_1E_{\alpha}(at^{\alpha})$$

$$E_{\alpha}(-at^{\alpha})(D^{\alpha}y - ay) = A_1E_{\alpha}(at^{\alpha})E_{\alpha}(-at^{\alpha})$$

$$D^{\alpha}[yE_{\alpha}(-at^{\alpha})] = A_1 = D^{\alpha}\left[\frac{A_1t^{\alpha}}{\Gamma(1+\alpha)}\right]$$

$$yE_{\alpha}(-at^{\alpha}) = \frac{A_1}{\Gamma(1+\alpha)} + B$$

$$y = \left(At^{\alpha} + B\right)E_{\alpha}(at^{\alpha}) \qquad \text{where} \qquad A = \frac{A_1}{\Gamma(1+\alpha)}$$

A and *B* are constants.

Thus the following theorem can be stated

Theorem 3: The fractional differential equation

$$D^{2\alpha}y - 2aD^{\alpha}y + a^2y = 0$$

has solution of the form

$$y = (At^{\alpha} + B)E_{\alpha}(at^{\alpha})$$

where A and B are constants.

The proof of the theorem is already explained in the previous arguments.

Theorem 4: Solution of the fractional differential equation

$$D^{2\alpha}y - 2aD^{\alpha}y + (a^2 + b^{\alpha})y = 0$$

is of the form

$$y = E_{\alpha}(at^{\alpha})[A\cos_{\alpha}(bt^{\alpha}) + B\sin_{\alpha}(bt^{\alpha})].$$

Proof: The given differential equation can be written in the following form

$$\begin{aligned} &((D^{\alpha} - a)^2 + b^2)y = 0 \\ &\text{or } (D^{\alpha} - a + ib)(D^{\alpha} - a - ib)y = 0 \end{aligned} \tag{10}$$

Using theorem 3 we get the solution of the fractional differential (10) can be written in the following form

$$y = A_1 E_{\alpha}((a + ib)t^{\alpha}) + B_1 E_{\alpha}((a - ib)t^{\alpha})$$

$$\begin{aligned} y &= A_1 E_{\alpha}(at^{\alpha})E_{\alpha}(ibt^{\alpha}) + B_1 E_{\alpha}(at^{\alpha})E_{\alpha}(-ibt^{\alpha}) \\ &= E_{\alpha}(at^{\alpha})[A_1\{\cos_{\alpha}(bt^{\alpha}) + i\sin_{\alpha}(bt^{\alpha})\} \\ &\quad + B_1\{\cos_{\alpha}(bt^{\alpha}) - i\sin_{\alpha}(bt^{\alpha})\}] \\ &= E_{\alpha}(at^{\alpha})[A\cos_{\alpha}(bt^{\alpha}) + iB\sin_{\alpha}(bt^{\alpha})] \end{aligned}$$

Where $A = A_1 + B_1$ and $B = A_1 - B_1$.

Thus we get useful results.

CONCLUSIONS

There are several methods to solve fractional differential equations, and the solution depends on the type of fractional derivative used. Here we develop an analytical method to find the solutions of linear fractional differential equation, composed by Jumarie fractional derivative in terms of one parameter Mittag-Leffler function. Some well known properties of Mittag-Leffler have been used to find solution of the fractional differential equations. The solutions obtained are similar as the solutions obtained usual calculus obtained in terms the exponential function. This conjugation with ordinary calculus when Jumarie type fractional derivative is used to compose the fractional differential equations is useful in several physical problems.

ACKNOWLEDGEMENT

Acknowledgments are to **Board of Research in Nuclear Science** (BRNS), Department of Atomic Energy Government of India for financial assistance received through BRNS research project no. 37(3)/14/46/2014-BRNS with BSC BRNS, title "Characterization of unreachable (Holderian) functions via Local Fractional Derivative and Deviation Function.

REFERENCES

1. de Oliveira, E. C., J. A. T. Machado. A review of definitions of fractional derivatives and Integral. Mathematical Problems in Engineering. Hindawi Publishing Corporation. 2014. 1-6.
2. Jumarie, G. Modified Riemann-Liouville derivative and fractional Taylor series of non-differentiable functions Further results, Computers and Mathematics with Applications, 2006. (51), 1367-1376.
3. G. Jumarie, "On the solution of the stochastic differential equation of exponential growth driven by fractional Brownian motion," AppliedMathematics Letters, vol. 18, no. 7, pp. 817-826, 2005.
4. G. Jumarie, "An approach to differential geometry of fractional order via modified Riemann-Liouville derivative,"Acta Mathematica Sinica, vol. 28, no. 9, pp. 1741-1768, 2012.
5. G. Jumarie, "On the derivative chain-rules in fractional calculus via fractional difference and their application to systems modelling," Central European Journal of Physics, vol. 11, no. 6, pp. 617-633, 2013.
6. X. J. Yang, Local Fractional Functional Analysis and Its Applications, Asian Academic Publisher Limited, Hong Kong, 2011.
7. X. J. Yang, Advanced Local Fractional Calculus and Its Applications, World Science, New York, NY, USA, 2012.
8. Miller KS, Ross B. An Introduction to the Fractional Calculus and Fractional Differential Equations.John Wiley & Sons, New York, NY, USA; 1993.
9. Das. S. Functional Fractional Calculus 2nd Edition, Springer-Verlag 2011.
10. Podlubny I. Fractional Differential Equations, Mathematics in Science and Engineering, Academic Press, San Diego, Calif, USA. 1999; 198.

11. M. Caputo, "Linear models of dissipation whose q is almost frequency independent-ii," Geophysical Journal of the Royal Astronomical Society, 1967. vol. 13, no. 5, pp. 529-539.

12. K M Kolwankar and A D. Gangal. Local fractional Fokker plank equation, Phys Rev Lett. 80 1998.

13. Abhay Parvate, A. D. Gangal. Calculus of fractals subset of real line: formulation-1; World Scientific, Fractals Vol. 17, 2009.

14. Abhay Parvate, Seema satin and A.D.Gangal. Calculus on a fractal curve in Rn arXiv:00906 oo76v1 3.6.2009; also in Pramana-J-Phys.

15. Abhay parvate, A. D. Gangal. Fractal differential equation and fractal time dynamic systems, Pramana-J-Phys, Vol 64, No. 3, 2005 pp 389-409.

16. E. Satin, Abhay Parvate, A. D. Gangal. Fokker-Plank Equation on Fractal Curves, Seema, Chaos Solitons & Fractals-52 2013, pp 30-35.

17. Erdelyi, A. Asymptotic expansions, Dover (1954).

18. Erdelyi, A. (Ed). Tables of Integral Transforms. vol. 1, McGraw-Hill, 1954.

19. Erdelyi, A. On some functional transformation Univ Potitec Torino 1950.

20. Mittag-Leffler, G. M.. Sur la nouvelle fonction Eα (x), C. R. Acad. Sci. Paris, (Ser. II) 137, 554-558 (1903).

21. Wiman, A. Uberden fundementalsatz in the theorie der funktionem Eα(z) Acta Math., 29. 1905. 191-201.

22. G. Jumarie. Fourier's transformation of fractional order via Mittag-Leffler function and modified Riemann-Liouville derivatives. J. Appl. Math. & informatics. 2008. 26. 1101-1121.

7

SIMILAR CONSTRUCTION METHOD OF THE BOUNDARY VALUE PROBLEM OF COMPOSITE SPHERICAL BESSEL EQUATIONS

Qiang Wang[1], Shunchu Li[1], Ming Hu[1], Dongdong Gui[2]

[1] Institute of Applied Mathematics, Xihua University, Chengdu, China

[2] Beijing Dong run ke Petroleum Technology Co,Ltd, Beijing, China

ABSTRACT

Aimed at the boundary value problem of composite spherical Bessel equations. Through introducing similar kernel function in the left (right) region, the similar structure of solution was obtained. By analyzing the structure of solution, the solution in the left region was assembled by coefficient of left boundary value condition and similar kernel function in the left region, the solution in the right region was assembled by similar kernel function in the left (right) region; coefficient of left boundary value condition; left guide functions and coefficient of joint conditions. Thus a simple method of solving the boundary value problem of composite spherical Bessel equations is put forward, namely similar construction method; This method provides a way for solving boundary value problem of composite differential equations.

INTRODUCTION

With the rapid development of science and technology, a lot of edge sciences are produced, and a lot of new-style differential equations are appeared; differential equation is also extending in chemistry and biology at the same time, and a lot of reaction diffusion equations are appeared; differential equation is also widely used in materials science[1,2], especially the composite material is widely used in industrial area, and some mathematical models of the composite material can be summed up in solving boundary value problem of composite differential equations, and which plays an important role in practical application.

In spherical coordinate system, Spherical Bessel equation is ordinary differential equation of radial component which is obtained by using the method of separation of variables to solve Laplace equation[3], it is common in well potential problem of the spherical square[4], its solution is called spherical Bessel function, and it is used to solve definite solution problem of heat conduction for spherically symmetric object[5,6], therefore solving the boundary value problem of spherical Bessel equation is of significant.

About the research of solution structure for the boundary value problem of differential equation, in 2004, similar construction theory of solution was put forward in reference[7], namely expression of solution for the boundary value problem of differential equation has the similar structure, in other words, it is expressed by the form of continued fraction, and this theory was demonstrated in many references [8-14]. Such as boundary value problem of homogeneous linear differential equations of second order; boundary value problem of Euler hyper-geometric equation and so on.

This paper studied the following the boundary value problem of composite spherical Bessel equations:

$$\begin{cases} x^2 z_1'' + 2xz_1' + \left[x^2 - n_1(n_1+1)\right]z_1 = 0(a<x<c) \\ x^2 z_2'' + 2xz_2' + \left[x^2 - n_2(n_2+1)\right]z_2 = 0(c<x<b) \\ \left[Mz_1 + (1+MN)z_1'\right]_{x=a} = E \\ z_1|_{x=c} = \lambda z_2|_{x=c}\ z_1'|_{x=c} = \mu z_2'|_{x=c} \\ \left[Pz_2 + Qz_2'\right]_{x=b} = 0 \end{cases} \tag{1}$$

Where: $a,b,c,E,M,N,P,Q,\lambda,\mu$ are real constant, $n_i(i=1,2)$ is positive constant including zero, and $E \neq 0$, $P^2+Q^2 \neq 0, 0<a<c<b$.

Aimed at the boundary value problem (1), in part one, three lemmas are given; in part two, similar structure theorem of solution is demonstrated, similar kernel function in the left and right region is introduced, and two corollaries are obtained; in part three, the concrete step of similar construction method is given; finally, superiority of similar construction method is illuminated by giving a example.

PRELIMINARY KNOWLEDGE

Lemma 1. [15] spherical Bessel equation

$$x^2y'' + 2xy' + \left[x^2 - n(n+1)\right]y = 0 \tag{2}$$

can be converted into the normal Bessel equation of $n+\frac{1}{2}$ order, namely

$$x^2\frac{d^2\xi}{dx^2} + x\frac{d\xi}{dx} + \left[x^2 - \left(n+\frac{1}{2}\right)^2\right]\xi = 0 \tag{3}$$

Proof: $x^2y'' + 2xy' + \left[x^2 - n(n+1)\right]y = 0$ is converted into the equation $x^2\frac{d^2\xi}{dx^2} + x\frac{d\xi}{dx} + \left[x^2 - \left(n+\frac{1}{2}\right)^2\right]\xi = 0$ by variable substitution $y = \sqrt{\frac{\pi}{2x}}\xi$.

Lemma 2. the general solution of spherical Bessel equation is

$$y = \sqrt{\frac{\pi}{2x}}\left[C_1Y_{n+\frac{1}{2}}(x) + C_2J_{n+\frac{1}{2}}(x)\right] \tag{4}$$

Where: C_1, C_2 are any constant, $J_{n+\frac{1}{2}}(), Y_{n+\frac{1}{2}}()$ are first and second class Bessel function of $n+\frac{1}{2}$ order respective.

Proof: Because the general solution [16] of the normal Bessel equation (3) is

$$\xi = C_1Y_{n+\frac{1}{2}}(x) + C_2J_{n+\frac{1}{2}}(x)$$

$y = \sqrt{\frac{\pi}{2x}}\xi = \sqrt{\frac{\pi}{2x}}\left[C_1Y_{n+\frac{1}{2}}(x) + C_2J_{n+\frac{1}{2}}(x)\right]$ is obtained by variable substitution $y = \sqrt{\frac{\pi}{2x}}\xi$.

Lemma 3. function of two variables is structured, namely

$$\psi_{m,n}(x,y) = Y_m(x)J_n(y) - J_m(x)Y_n(y) \tag{5}$$

Where: m,n are constant, $J_v(\cdot), Y_v(\cdot)$ are first and second Bessel function of v order respective.

Therefore the following formula is easily obtained:

$$\frac{\partial}{\partial x}\psi_{v,v}(x,y) = \frac{v}{x}\psi_{v,v}(x,y) - \psi_{v+1,v}(x,y) \tag{6}$$

$$\frac{\partial}{\partial y}\psi_{v,v}(x,y)=\frac{v}{y}\psi_{v,v}(x,y)-\psi_{v,v+1}(x,y) \tag{7}$$

$$=\frac{v^2}{xy}\psi_{v,v}(x,y)-\frac{v}{x}\psi_{v,v+1}(x,y)-\frac{v}{y}\psi_{v+1,v}(x,y,t)$$
$$+\psi_{v+1,v+1}(x,y) \tag{8}$$

Proof: According to the differential property of Bessel function[16]:

$$\frac{d}{dt}J_v(t)=\frac{v}{t}J_v(t)-J_{v+1}(t),$$
$$\frac{d}{dt}Y_v(t)=\frac{v}{t}Y_v(t)-Y_{v+1}(t).$$

We can obtain

$$\frac{\partial}{\partial x}\psi_{v,v}(x,y)=\frac{d}{dx}\left[Y_v(x)J_v(y)-J_v(x)Y_v(y)\right]$$
$$=\left[\frac{v}{x}Y_v(x)-Y_{v+1}(x)\right]J_v(y)-\left[\frac{v}{x}J_v(x)-J_{v+1}(x)\right]Y_v(y)$$
$$=\frac{v}{x}\psi_{v,v}(x,y)-\psi_{v+1,v}(x,y)$$

By a similar way, (7) and (8) can be demonstrated.

The guide functions are structured by two linearly independent solutions $\sqrt{\frac{\pi}{2x}}Y_{n_i+\frac{1}{2}}(x),\sqrt{\frac{\pi}{2x}}J_{n_i+\frac{1}{2}}(x)$ of spherical Bessel equation, namely

$$\begin{aligned}\phi_{0,0}^{i}(x,\xi)&=\sqrt{\frac{\pi}{2x}}Y_{n_i+\frac{1}{2}}(x)\cdot\sqrt{\frac{\pi}{2\xi}}J_{n_i+\frac{1}{2}}(\xi)\\&-\sqrt{\frac{\pi}{2x}}J_{n_i+\frac{1}{2}}(x)\cdot\sqrt{\frac{\pi}{2\xi}}Y_{n_i+\frac{1}{2}}(\xi)\\&=\frac{\pi}{2}\sqrt{\frac{1}{x\xi}}\psi_{n_i+\frac{1}{2},n_i+\frac{1}{2}}(x,\xi)\end{aligned} \tag{9}$$

$$\begin{aligned}\phi_{1,0}^{i}(x,\xi) &= \frac{\partial}{\partial x}\phi_{0,0}^{i}(x,\xi) \\ &= \frac{\pi}{2}\sqrt{\frac{1}{x\xi}}\left[\frac{n_i}{x}\psi_{n_i+\frac{1}{2},n_i+\frac{1}{2}}(x,\xi)-\psi_{n_i+\frac{3}{2},n_i+\frac{1}{2}}(x,\xi)\right]\end{aligned} \tag{10}$$

$$\begin{aligned}\phi_{0,1}^{i}(x,\xi) &= \frac{\partial}{\partial \xi}\phi_{0,0}^{i}(x,\xi) \\ &= \frac{\pi}{2}\sqrt{\frac{1}{x\xi}}\left[\frac{n_i}{\xi}\psi_{n_i+\frac{1}{2},n_i+\frac{1}{2}}(x,\xi)-\psi_{n_i+\frac{1}{2},n_i+\frac{3}{2}}(x,\xi)\right]\end{aligned} \tag{11}$$

$$\begin{aligned}\phi_{1,1}^{i}(x,\xi) &= \frac{\partial^2}{\partial x\partial \xi}\phi_{0,0}^{i}(x,\xi) = \frac{\partial^2}{\partial \xi\partial x}\phi_{0,0}^{i}(x,\xi) \\ &= \frac{\pi n_i^2}{2}\sqrt{\frac{1}{x\xi}}\frac{1}{x\xi}\psi_{n_i+\frac{1}{2},n_i+\frac{1}{2}}(x,\xi) \\ &\quad -\frac{\pi n_i}{2}\sqrt{\frac{1}{x\xi}}\frac{1}{x}\psi_{n_i+\frac{1}{2},n_i+\frac{3}{2}}(x,\xi) \\ &\quad -\frac{\pi n_i}{2}\sqrt{\frac{1}{x\xi}}\frac{1}{\xi}\psi_{n_i+\frac{3}{2},n_i+\frac{1}{2}}(x,\xi) \\ &\quad +\frac{\pi}{2}\sqrt{\frac{1}{x\xi}}\psi_{n_i+\frac{3}{2},n_i+\frac{3}{2}}(x,\xi)\end{aligned} \tag{12}$$

Where: $i = 1, 2$.

THE MAIN THEOREM AND ITS PROOF

Theorem If the boundary value problem (1) of composite spherical Bessel equations has unique solution, its solution in the left region ($a < x < c$) is

$$z_1 = E\cdot\frac{1}{M+\frac{1}{N+\varphi(a)}}\cdot\frac{1}{N+\varphi(a)}\cdot\varphi(x) \tag{13}$$

its solution in the right region ($c < x < b$) is

$$z_2 = E \cdot \frac{1}{M + \frac{1}{N+\varphi(a)}} \cdot \frac{1}{N+\varphi(a)} \tag{14}$$

$$\cdot \frac{\phi_{0,1}^{1}(c,c)}{\lambda \varphi^{*}(c)\phi_{1,1}^{1}(a,c) - \mu \phi_{1,0}^{1}(a,c)} \cdot \varphi^{*}(x)$$

Where: $\varphi(x)$ is called similar kernel function in the left region, namely

$$\varphi(x) = \frac{\lambda \varphi^{*}(c)\phi_{0,1}^{1}(x,c) - \mu \phi_{0,0}^{1}(x,c)}{\lambda \varphi^{*}(c)\phi_{1,1}^{1}(a,c) - \mu \phi_{1,0}^{1}(a,c)}, (a < x < c) \tag{15}$$

$\varphi^{*}(x)$ is called similar kernel function in the right region, namely

$$\varphi^{*}(x) = \frac{P\phi_{0,0}^{2}(x,b) + Q\phi_{0,1}^{2}(x,b)}{P\phi_{1,0}^{2}(c,b) + Q\phi_{1,1}^{2}(c,b)}, (c < x < b) \tag{16}$$

Proof: According to lemma 2, the general solution of the definite solution equation for the boundary value problem (1) is

$$z_i(x) = C_i^1 \sqrt{\frac{\pi}{2x}} Y_{n_i + \frac{1}{2}}(x) + C_i^2 \sqrt{\frac{\pi}{2x}} J_{n_i + \frac{1}{2}}(x), (i = 1,2) \tag{17}$$

Where: $C_i^1, C_i^2 \, (i = 1,2)$ are any constant; when $i = 1$, it indicates left region ($a < x < c$); when $i = 2$, it indicates right region ($c < x < b$).

Taking the derivative of $z_i(x)$, namely

$$z_i^{'}(x) = \frac{d}{dx}\left[C_i^1 \sqrt{\frac{\pi}{2x}} Y_{n_i + \frac{1}{2}}(x) + C_i^2 \sqrt{\frac{\pi}{2x}} J_{n_i + \frac{1}{2}}(x) \right]$$

$$= \sqrt{\frac{\pi}{2}} x^{-\frac{3}{2}} n_i \left[C_i^1 Y_{n_i + \frac{1}{2}}(x) + C_i^2 J_{n_i + \frac{1}{2}}(x) \right]$$

$$- \sqrt{\frac{\pi}{2x}} \left[C_i^1 Y_{n_i + \frac{3}{2}}(x) + C_i^2 J_{n_i + \frac{3}{2}}(x) \right]$$

The following formulas are obtained by putting the formula (17) plug into the boundary condition of the boundary value problem (1).

$$\begin{bmatrix} \left(M\sqrt{\frac{\pi}{2a}} + (1+MN)\sqrt{\frac{\pi}{2}}a^{-\frac{3}{2}}n_1 \right) Y_{n_1+\frac{1}{2}}(a) \\ -(1+MN)\sqrt{\frac{\pi}{2a}}Y_{n_1+\frac{3}{2}}(a) \end{bmatrix} C_1^1 + \begin{bmatrix} \left(M\sqrt{\frac{\pi}{2a}} + (1+MN)\sqrt{\frac{\pi}{2}}a^{-\frac{3}{2}}n_1 \right) J_{n_1+\frac{1}{2}}(a) \\ -(1+MN)\sqrt{\frac{\pi}{2a}}J_{n_1+\frac{3}{2}}(a) \end{bmatrix} C_1^2 = E \tag{18}$$

$$\begin{aligned} &\sqrt{\frac{\pi}{2c}}Y_{n_1+\frac{1}{2}}(c)C_1^1 + \sqrt{\frac{\pi}{2c}}J_{n_1+\frac{1}{2}}(c)C_1^2 \\ &-\lambda\sqrt{\frac{\pi}{2c}}Y_{n_2+\frac{1}{2}}(c)C_2^1 - \lambda\sqrt{\frac{\pi}{2c}}J_{n_2+\frac{1}{2}}(c)C_2^2 = 0 \end{aligned} \tag{19}$$

$$\begin{aligned} &\left[\sqrt{\frac{\pi}{2}}c^{-\frac{3}{2}}n_1 Y_{n_1+\frac{1}{2}}(c) - \sqrt{\frac{\pi}{2c}}Y_{n_1+\frac{3}{2}}(c) \right] C_1^1 \\ &+\left[\sqrt{\frac{\pi}{2}}c^{-\frac{3}{2}}n_1 J_{n_1+\frac{1}{2}}(c) - \sqrt{\frac{\pi}{2c}}J_{n_1+\frac{3}{2}}(c) \right] C_1^2 - \\ &\left[\mu\sqrt{\frac{\pi}{2}}c^{-\frac{3}{2}}n_2 Y_{n_2+\frac{1}{2}}(c) - \mu\sqrt{\frac{\pi}{2c}}Y_{n_2+\frac{3}{2}}(c) \right] C_2^1 \\ &-\left[\mu\sqrt{\frac{\pi}{2}}c^{-\frac{3}{2}}n_2 J_{n_2+\frac{1}{2}}(c) - \mu\sqrt{\frac{\pi}{2c}}J_{n_2+\frac{3}{2}}(c) \right] C_2^2 = 0 \end{aligned} \tag{20}$$

$$\begin{bmatrix} P\sqrt{\frac{\pi}{2b}}Y_{n_2+\frac{1}{2}}(b) + Q\sqrt{\frac{\pi}{2}}b^{-\frac{3}{2}}n_2 Y_{n_2+\frac{1}{2}}(b) \\ -Q\sqrt{\frac{\pi}{2b}}Y_{n_2+\frac{3}{2}}(b) \end{bmatrix} C_2^1 + \tag{21}$$

$$\begin{bmatrix} P\sqrt{\frac{\pi}{2b}}J_{n_2+\frac{1}{2}}(b)+Q\sqrt{\frac{\pi}{2}}b^{-\frac{3}{2}}n_2J_{n_2+\frac{1}{2}}(b) \\ -Q\sqrt{\frac{\pi}{2b}}J_{n_2+\frac{3}{2}}(b) \end{bmatrix} C_2^2=0$$

Because the boundary value problem (1) has unique solution, determinant Δ of coefficients $C_1^1, C_1^2, C_2^1, C_2^2$ for equations (18)-(21) is not equal to zero, according to formulas (9)-(12), we can obtain

$$\begin{aligned} \Delta = {} & \lambda\left[P\phi_{0,0}^2(c,b)+Q\phi_{0,1}^2(c,b)\right] \\ & \cdot\left[M\phi_{0,1}^1(a,c)+(1+MN)\phi_{1,1}^1(a,c)\right] \\ & -\mu\left[P\phi_{1,0}^2(c,b)+Q\phi_{1,1}^2(c,b)\right] \\ & \cdot\left[M\phi_{0,0}^1(a,c)+(1+MN)\phi_{1,0}^1(a,c)\right] \end{aligned} \tag{22}$$

and then according to grammer rule, we can obtain value of coefficients $C_1^1, C_1^2, C_2^1, C_2^2$, namely

$$C_1^1=\frac{E}{\Delta}\left\{\begin{array}{l} \lambda\left[P\phi_{0,0}^2(c,b)+Q\phi_{0,1}^2(c,b)\right] \\ \cdot\left[\sqrt{\frac{\pi}{2c}}\left(\frac{n_1}{c}J_{n_1+\frac{1}{2}}(c)-J_{n_1+\frac{3}{2}}(c)\right)\right] \end{array}\right. \tag{23}$$

$$-\mu\left[P\phi_{1,0}^2(c,b)+Q\phi_{1,1}^2(c,b)\right]\sqrt{\frac{\pi}{2c}}J_{n_1+\frac{1}{2}}(c)$$

$$C_1^2=\frac{-E}{\Delta}\left\{\begin{array}{l} \lambda\left[P\phi_{0,0}^2(c,b)+Q\phi_{0,1}^2(c,b)\right] \\ \cdot\left[\sqrt{\frac{\pi}{2c}}\begin{pmatrix} \frac{n_1}{c}Y_{n_1+\frac{1}{2}}(c) \\ -Y_{n_1+\frac{3}{2}}(c) \end{pmatrix}\right] \end{array}\right. \tag{24}$$

$$-\mu\left[P\phi_{1,0}^{2}(c,b)+Q\phi_{1,1}^{2}(c,b)\right]\sqrt{\frac{\pi}{2c}}Y_{n_1+\frac{1}{2}}(c)\Bigg\}$$

$$C_2^1=\frac{E}{\Delta}\left\{\begin{array}{l}P\sqrt{\frac{\pi}{2b}}J_{n_2+\frac{1}{2}}(b)\\+Q\left[\sqrt{\frac{\pi}{2b}}\left(\begin{array}{l}\frac{n_2}{b}J_{n_2+\frac{1}{2}}(b)\\-J_{n_2+\frac{3}{2}}(b)\end{array}\right)\right]\end{array}\right\}\phi_{0,1}^{1}(c,c) \tag{25}$$

$$C_2^2=\frac{-E}{\Delta}\left\{\begin{array}{l}P\sqrt{\frac{\pi}{2b}}Y_{n_2+\frac{1}{2}}(b)\\+Q\left[\sqrt{\frac{\pi}{2b}}\left(\begin{array}{l}\frac{n_2}{b}Y_{n_2+\frac{1}{2}}(b)\\-Y_{n_2+\frac{3}{2}}(b)\end{array}\right)\right]\end{array}\right\}\phi_{0,1}^{1}(c,c) \tag{26}$$

By putting the formula (23)-(26) plug into the formula (17) and using the similar kernel function in the left and right region (15);(16), the solution in the left and right region of boundary value problem (1) is obtained, namely (13);(14).

According to the theorem, it is easy to obtain the following corollary.

Corollary 1. If right boundary value condition of the boundary value problem (1) is $z_2(b)=0$, namely $Q=0,\ P\neq 0$, the corresponding similar kernel function in the right region is

$$\varphi^*(x)=\frac{\phi_{0,0}^{2}(x,b)}{\phi_{1,0}^{2}(c,b)}.$$

Corollary 2. If right boundary value condition of the boundary value problem (1) is $z_2'(b)=0$, namely $P=0, Q\neq 0$, the corresponding similar kernel function in the right region is

$$\varphi^*(x)=\frac{\phi_{0,1}^{2}(x,b)}{\phi_{1,1}^{2}(c,b)}$$

STEP OF SIMILAR CONSTRUCTION METHOD

According to above the solving process for the boundary value problem (1), by analysis and summary, it is easy to obtain concrete step of similar construction method, which solving the boundary value (1).

Step 1. The function of two variables (5) is constructed, and then the guide functions (9)-(12) are constructed by using the two linearly independent solutions $\sqrt{\frac{\pi}{2x}}Y_{n_i+\frac{1}{2}}(x)$、$\sqrt{\frac{\pi}{2x}}J_{n_i+\frac{1}{2}}(x)(i=1,2)$ of the definite solution equation for the boundary value problem (1).

Step 2. According to the guide functions in the right region $\phi_{0,0}^{2}(x,\xi)$、$\phi_{0,1}^{2}(x,\xi)$、$\phi_{1,0}^{2}(x,\xi)$、$\phi_{1,1}^{2}(x,\xi)$ and the coefficients P、Q of the right boundary value condition $\left[Pz_2+Qz_2'\right]_{x=b}=0$, the similar kernel function in the right region $\varphi^{*}(x)$ is assembled by using the formula (16); According to the guide functions in the left region $\phi_{0,0}^{1}(x,\xi)$、$\phi_{0,1}^{1}(x,\xi)$、$\phi_{1,0}^{1}(x,\xi)$、$\phi_{1,1}^{1}(x,\xi)$;$\varphi^{*}(c)$ and the coefficients λ、μ of the connective condition $z_1|_{x=c}=\lambda z_2|_{x=c}, z_1'|_{x=c}=\mu z_2'|_{x=c}$, the similar kernel function in the left region $\varphi(x)$ is assembled by using the formula (15).

Step 3. According to the coefficients E、M、N of the left boundary value condition $\left[Mz_1+(1+MN)z_1'\right]_{x=a}=E$ and $\varphi(x)$、$\varphi(a)$, the solution in the left region of the boundary value problem (1) is assembled by using the formula (13); According to the coefficients E、M、N of the left boundary value condition $\left[Mz_1+(1+MN)z_1'\right]_{x=a}=E$ and $\varphi(a)$, the guide functions in the left region $\phi_{0,1}^{1}(x,\xi)$、$\phi_{1,0}^{1}(x,\xi)$、$\phi_{1,1}^{1}(x,\xi)$, the coefficients λ、μ of joint conditions $z_1|_{x=c}=\lambda z_2|_{x=c}, z_1'|_{x=c}=\mu z_2'|_{x=c}$ and the similar kernel function in the right region $\varphi^{*}(x)$, the solution in the right region is assembled by using the formula (14) for the boundary value problem (1).

EXAMPLE

The following boundary value problem is solved by using the above step of similar construction method.

$$\begin{cases} x^2 z_1'' + 2x z_1' + \left(x^2 - 2\right) z_1 = 0 \left(1 < x < 10\right) \\ x^2 z_2'' + 2x z_2' + \left(x^2 - 6\right) z_2 = 0 \left(10 < x < 20\right) \\ \left(z_1 + 2 z_1'\right)_{x=1} = 1 \\ z_1 \big|_{x=10} = z_2 \big|_{x=10} \\ z_1' \big|_{x=10} = z_2' \big|_{x=10} \\ \left(z_2 + z_2'\right)_{x=20} = 0 \end{cases} \tag{27}$$

Where: In this case $a = 1, b = 20, c = 10, M = N = E = P = Q = \lambda = \mu = 1,$ $n_1 = 1, n_2 = 2.$

Step 1. The function of two variables and guide functions are constructed, namely $\psi_{\frac{3}{2},\frac{3}{2}}(x,\xi) = Y_{\frac{3}{2}}(x) J_{\frac{3}{2}}(\xi) - J_{\frac{3}{2}}(x) Y_{\frac{3}{2}}(\xi)$ and $\psi_{\frac{5}{2},\frac{5}{2}}(x,\xi) = Y_{\frac{5}{2}}(x) J_{\frac{5}{2}}(\xi) - J_{\frac{5}{2}}(x) Y_{\frac{5}{2}}(\xi)$ are called function of two variables, the following functions are called guide function.

$$\begin{aligned} \phi_{0,0}^1(x,\xi) &= \sqrt{\frac{\pi}{2x}} Y_{\frac{3}{2}}(x) \cdot \sqrt{\frac{\pi}{2\xi}} J_{\frac{3}{2}}(\xi) \\ &\quad - \sqrt{\frac{\pi}{2x}} J_{\frac{3}{2}}(x) \cdot \sqrt{\frac{\pi}{2\xi}} Y_{\frac{3}{2}}(\xi) \\ &= \frac{\pi}{2} \sqrt{\frac{1}{x\xi}} \psi_{\frac{3}{2},\frac{3}{2}}(x,\xi) \end{aligned}$$

$$\begin{aligned} \phi_{0,0}^2(x,\xi) &= \sqrt{\frac{\pi}{2x}} Y_{\frac{5}{2}}(x) \cdot \sqrt{\frac{\pi}{2\xi}} J_{\frac{5}{2}}(\xi) \\ &\quad - \sqrt{\frac{\pi}{2x}} J_{\frac{5}{2}}(x) \cdot \sqrt{\frac{\pi}{2\xi}} Y_{\frac{5}{2}}(\xi) \\ &= \frac{\pi}{2} \sqrt{\frac{1}{x\xi}} \psi_{\frac{5}{2},\frac{5}{2}}(x,\xi), \end{aligned}$$

$$\phi_{1,0}^{1}(x,\xi)=\frac{\partial}{\partial x}\phi_{0,0}^{1}(x,\xi)$$
$$=\frac{\pi}{2}\sqrt{\frac{1}{x\xi}}\left[\tfrac{1}{x}\psi_{\frac{3}{2},\frac{3}{2}}(x,\xi)-\psi_{\frac{5}{2},\frac{3}{2}}(x,\xi)\right]$$

$$\phi_{1,0}^{2}(x,\xi)=\frac{\partial}{\partial x}\phi_{0,0}^{2}(x,\xi)$$
$$=\frac{\pi}{2}\sqrt{\frac{1}{x\xi}}\left[\tfrac{2}{x}\psi_{\frac{5}{2},\frac{5}{2}}(x,\xi)-\psi_{\frac{7}{2},\frac{5}{2}}(x,\xi)\right]$$

$$\phi_{0,1}^{1}(x,\xi)=\frac{\partial}{\partial \xi}\phi_{0,0}^{1}(x,\xi)$$
$$=\frac{\pi}{2}\sqrt{\frac{1}{x\xi}}\left[\tfrac{1}{\xi}\psi_{\frac{3}{2},\frac{3}{2}}(x,\xi)-\psi_{\frac{3}{2},\frac{5}{2}}(x,\xi)\right]$$

$$\phi_{0,1}^{2}(x,\xi)=\frac{\partial}{\partial \xi}\phi_{0,0}^{2}(x,\xi)$$
$$=\frac{\pi}{2}\sqrt{\frac{1}{x\xi}}\left[\tfrac{2}{\xi}\psi_{\frac{5}{2},\frac{5}{2}}(x,\xi)-\psi_{\frac{5}{2},\frac{7}{2}}(x,\xi)\right]$$

$$\phi_{1,1}^{1}(x,\xi)=\frac{\partial^2}{\partial x\partial \xi}\phi_{0,0}^{1}(x,\xi)=\frac{\partial^2}{\partial \xi\partial x}\phi_{0,0}^{1}(x,\xi)$$
$$=\frac{\pi}{2}\sqrt{\frac{1}{x\xi}}\frac{1}{x\xi}\psi_{\frac{3}{2},\frac{3}{2}}(x,\xi)$$
$$-\frac{\pi}{2}\sqrt{\frac{1}{x\xi}}\frac{1}{x}\psi_{\frac{3}{2},\frac{5}{2}}(x,\xi)$$
$$-\frac{\pi}{2}\sqrt{\frac{1}{x\xi}}\frac{1}{\xi}\psi_{\frac{5}{2},\frac{3}{2}}(x,\xi)$$
$$+\frac{\pi}{2}\sqrt{\frac{1}{x\xi}}\psi_{\frac{5}{2},\frac{5}{2}}(x,\xi)$$

$$\phi_{1,1}^{2}(x,\xi)=\frac{\partial^{2}}{\partial x\partial\xi}\phi_{0,0}^{2}(x,\xi)=\frac{\partial^{2}}{\partial\xi\partial x}\phi_{0,0}^{2}(x,\xi)$$

$$=2\pi\sqrt{\frac{1}{x\xi}}\frac{1}{x\xi}\psi_{\frac{5}{2},\frac{5}{2}}(x,\xi)$$

$$-\pi\sqrt{\frac{1}{x\xi}}\frac{1}{x}\psi_{\frac{5}{2},\frac{7}{2}}(x,\xi)$$

$$-\pi\sqrt{\frac{1}{x\xi}}\frac{1}{\xi}\psi_{\frac{7}{2},\frac{5}{2}}(x,\xi)$$

$$+\frac{\pi}{2}\sqrt{\frac{1}{x\xi}}\psi_{\frac{7}{2},\frac{7}{2}}(x,\xi)$$

Step 2. The similar kernel functions are constructed, namely $\varphi^{*}(x)=\dfrac{\phi_{0,0}^{2}(x,20)+\phi_{0,1}^{2}(x,20)}{\phi_{1,0}^{2}(10,20)+\phi_{1,1}^{2}(10,20)}$ is called similar kernel function in the right region of the boundary value problem (27).

$\varphi(x)=\dfrac{\varphi^{*}(10)\phi_{0,1}^{1}(x,10)-\phi_{0,0}^{1}(x,10)}{\varphi^{*}(10)\phi_{1,1}^{1}(1,10)-\phi_{1,0}^{1}(1,10)}$ is called similar kernel in the left region of the boundary value problem (27) and $\varphi(1)=\dfrac{\varphi^{*}(10)\phi_{0,1}^{1}(1,10)-\phi_{0,0}^{1}(1,10)}{\varphi^{*}(10)\phi_{1,1}^{1}(1,10)-\phi_{1,0}^{1}(1,10)}$.

Step 3. The solution is constructed, namely $z_1=\dfrac{1}{1+\dfrac{1}{1+\varphi(1)}}\cdot\dfrac{1}{1+\varphi(1)}\cdot\varphi(x)$ $(1<x<10)$ is called solution in the left region of the boundary value problem (27).

$$z_2=\frac{1}{1+\dfrac{1}{1+\varphi(1)}}\cdot\frac{1}{1+\varphi(1)}\cdot\frac{\phi_{0,1}^{1}(10,10)}{\varphi^{*}(10)\phi_{1,1}^{1}(1,10)-\phi_{1,0}^{1}(1,10)}\cdot\varphi^{*}(x),(10<x<20)$$

is called solution in the right region of the boundary value problem (27).

According to the concrete step cf similar construction method, the changing curve of the solution in the designated area is drawn by using *Matlab* for the boundary value problem (27), namely figure 1

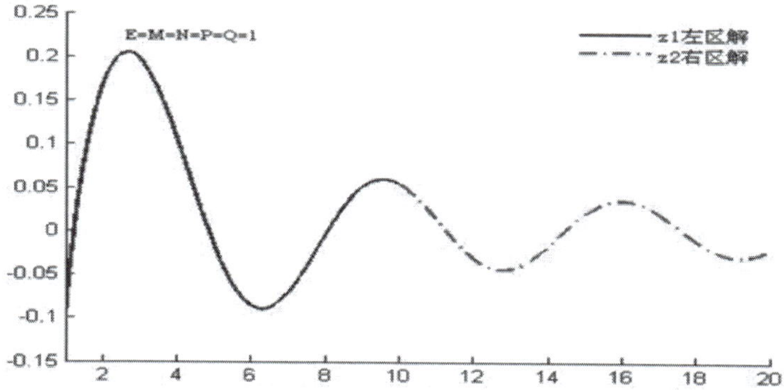

FIGURE 1. Curve of solution to the boundary value problem (27)

CONCLUSION

(1) Aimed at the solution (13) and (14) of the boundary value problem (1), it is expressed by continued fraction in form, namely similar structure.

(2) For solving process of the boundary value problem (1), similar construction method that is put forward can solve its solution, and this method is simple and effective, the step of this method plays an important role in drawing the changing curve of solution of the corresponding boundary value problem for men of mathematics.

(3) For the boundary value problem (1), when the coefficient of right boundary value condition changes, it is merely changing similar kernel function, but expression form of the solution is not changing.

(4) Observing the formula (13) and (14), the following relational expression is obtained, namely $\left[z_1(x)+Nz_1'(x)\right]_{x=a}=\dfrac{E}{M+\dfrac{1}{N+\varphi(a)}}$.

ACKNOWLEDGMENT

This Project is supported by the Scientific Research Fund of Sichuan Provincial Education Department of China under Grant No.12ZA164.

REFERENCES

1. Yiting Wang, Ping Wang, Xun Wang. The application of the differential equation in the studies of materials (in Chinese). Academy, 2011, 2 (4): 21-22.

2. Xiufen Wang, Zhihong Ma, Zhimin Mu. Study of the temperature control material force deformation Based on the differential equation model (in Chinese). Coal Technology, 2010, 24 (5): 208-210.

3. Shanjie zhang. Vector analysis; cylindrical function and spherical function. Nanjing: Nanjing University Press, 2011.
4. Yongde Zhang. Quantum mechanics. Beijing: Science Press, 2005.
5. Ying Gao, Hanwei Yi. Sula Baiya, ect. Computation of energy levels of I2 molecule by spherical Bessel function expansion method (in Chinese). Journal of Changchun University of science and technology, 2009, 32 (3): 511-512.
6. Xiaozeng Wang, Hui Zeng, Jiuhong Yang. Analysis on steel ball surface temperature change during heat treatment process (in Chinese). Hot Working Technology, 2012, 41 (4): 187-189 (-193).
7. Shunchu Li, Minhui Jia. The Formal Similarity of Solutions on the Class of Differential Equation. Journal of UEST of China, 2004, 33 (Supp.): 95-98.
8. Pengshe Zheng, Shunchu Li, Weibing Zhu. The similar structure of pressure distribution in the double porosity composite reservoir (in Chinese). Drilling & Production Technology, 2008, 31 (4): 80-81.
9. Shunchu Li. The similar structure of solution to the boundary value problem for second-order linear homogeneous differential equations (in Chinese). Journal of Xihua University (Natural Science Edition), 2009, 28 (5): 40-41 (to 90).
10. Shunchu Li. Preliminary exploration and prospects of the similar structure of solutions of differential equations (in Chinese). Journal of Xihua University (Natural Science Edition), 2010, 29 (2): 223-226 (to 238).
11. Yibin Tian, Shunchu Li, Quanyong Li, etc. The similar structure of Tschebyscheff equation on boundary value problem (in Chinese). Sichuan Ordnance Journal, 2011, 32 (1): 155-156.
12. Xuxia Xiao, Li Shunchu Li. The similar structure of solutions for the boundary value problem of Euler's Hypergeometric differential equation (in Chinese). Journal of Inner Mongolia Normal University (Natural Science Edition), 2012, 41 (6): 597-600 (-603)
13. Quanyong Li, Shunchu Li, Ke Li, etc. Study of the seepage model in dual-porosity composite media reservoir based on similar structure of solution (in Chinese). Drilling & Production Technology, 2012, 35 (2): 54-56.
14. Xitao Bao, Shunchu Li, ZhiJian Liao. Similar constructive method of solution for the boundary value problem of Euler Hyper-geometric differential equation (in Chinese). Journal of Southwest University of Science and Technology, 2012, 27 (4): 101-105.
15. Shunchu Li, Bingguang Huang. Laplace transform and Bessel functions and the theoretical basis of well test analysis. Beijing: Petroleum Industry Press, 2000.
16. Shishi Liu, Shida Liu. Special function. Beijing: China Meteorological Press, 2002.

8

COMPOSITION OPERATORS ON LORENTZ-KARAMATA-BOCHNER SPACES

Anuradha Gupta[1], Neha Bhatia[2,*]

[1] Department of Mathematics, Delhi College of Arts and Commerce, University of Delhi, Delhi-110023, India

[2] Department of Mathematics, University of Delhi, Delhi -110007, India

ABSTRACT

In this paper, study of the composition operators on Lorentz-Karamata-Bochner spaces and characterization of the properties like boundedness, closedness and essential range of these operators on the space has been made.

INTRODUCTION

Let $(\Omega, \mathcal{A}, \mu)$ be a σ-finite measure space. A measurable transformation T is said to be non-singular if $\mu\left(T^{-1}(A)\right) = 0$ 0 whenever $\mu(A) = 0$ for every $A \in \mathcal{A}$.

If T is non-singular, then we say that μT^{-1} is absolutely continuous with respect to μ. Hence, by Radon-Nikodym theorem there exists a unique non-negative essentially bounded function f_T such that $\mu\left(T^{-1}\right)(A) = \int_A f_T d\mu$ for $A \in \mathcal{A}$.

Let f be any complex-valued measurable function For $s \geq 0$, the *distribution function* μ_f of f is defined as

$$\mu_f(s) = \mu\{\omega \in \Omega : |f(\omega)| > s\}$$

The *non-increasing rearrangement* f^* of f is defined as

$$f^*(t) = \inf\{s > 0 : \mu_f(s) \leq t\}, \textit{ for all } t \geq 0$$

The *maximal (average) operator* is given by

$$f^{**}(t) = \frac{1}{t}\int_0^t f^*(s)\,ds$$

One can refer to [4] for the properties of these functions.

Definition 1. *A positive and Lebesgue measurable function b is said to be slowly varying (s.v.) on $(0, \infty)$ if for each $\epsilon > 0$, $t\epsilon b(t)$ is equivalent to a non-decreasing function and $t\epsilon b(t)$ is equivalent to a non-increasing function on $(0, \infty)$.*

Given a s.v. function b on $(0, \infty)$, we denote by γ_b the positive function defined by

$$\gamma_b(t) = b\left(\max\left\{t, \frac{1}{t}\right\}\right)$$

For various properties of slowly varying function we can refer to [4,10].

For $1 < p < \infty$, $1 \leq q < \infty$ and for a measurable function f *on* Ω, define

$$\| f \|_{p,q;\alpha} = \left\| t^{\frac{1}{p}-\frac{1}{q}} \gamma_b(t) f^{**}(t) \right\|_{q;(0,\infty)}$$

$$= \left(\int_0^\infty \left(t^{\frac{1}{p}-\frac{1}{q}} \gamma_b(t) f^{**}(t) \right)^q dt \right)^{\frac{1}{q}}$$

The Lorentz-Karamata space $L_{p,q;b}$ b introduced in [4] is the set of all measurable functions f on Ω, such that

$$\|f\|_{p,q;b} < \infty$$

Let $f : \Omega \to X$ be a strongly measurable function on a Banach space X. Define a function $\|f\|$ as

$$\|f\|(w) = \|f(w)\|$$

for all $\omega \in \Omega$. Then the Lorentz-Karamata-Bochner space $L_{p,q;b}(\Omega, X)$is a rearrangement invariant-Bochner space for $p, q \in (0, \infty)$ where the norm is given as

$$\|f\|_{p,q;b} = \left\| t^{\frac{1}{p}-\frac{1}{q}} \gamma_b(t) \|f^{**}\|(t) \right\|_{q;(0,\infty)}$$

The Lorentz-Karamata space $L_{p,q;b}$ is a Banach space and we still have the density of simple functions in it and its dual is

$$L^*_{p,q;b}(\Omega, X) = L_{p',q';b^{-1}}(\omega, X^*)$$

where X^* has the Radon-Nikodym property. For every $g \in L_{p',q';b^{-1}}$, we can find a bounded linear functional $F_g \in (L_{p,q;b})^* = L_{p',q';b^{-1}}$ defined as

$$F_g(f) = \int f g d\mu$$

for all $f \in L_{p,q;b}(\Omega, X)$. For each $g \in L_{p',q';b^{-1}}(\omega, X^*)$,there exists a unique $T^{-1}(\mathcal{A})$ measurable function E(g) such that

$$\int f g d\mu = \int f E(g) d\mu$$

for each $T^{-1}(\mathcal{A})$ measurable function f for which the left integral exists. E(g) is called the conditional expectation [11] of g with respect to $T^{-1}(\mathcal{A})$. The operator PT defined as

$$P_T g = f_T . E(g) \circ T^{-1}$$

is called Frobenius Perron and fT is the Radon-Nikodym derivative of μT^{-1} with respect to μ It satisfies the property

$$E(g) \circ T^{-1} = f \text{ if and only if } E(g) = f \circ T$$

Let T be a non-singular measurable transformation on Ω then the composition operator CT from $L_{p,q;b}(\Omega, X)$ into the space of strongly measurable functions $L(\Omega, X)$ is given by

$$(C_T f)(\omega) = f(T(\omega))$$

for all is $\omega \in \Omega$. An operator T is called Fredholm if R(T) is closed, dim $N(T) < \infty$ and dim $N(T^*) < \infty$ where R(T), N(T) and $N(T^*)$ denote the range, kernel and cokernel of T. $B(X)$ denotes the space of all bounded linear operators on X. Multiplication operators on this space are already studied in [5] and on different spaces in [1,2,3,6,7,8,9,12]. In this paper, we discuss about the composition operators on the Lorentz-Karamata-Bochner space and study its various properties like boundedness, closedness and compactness.

COMPOSITION OPERATORS

Theorem 2.1. *A non-singular transformation* $T : \Omega \to \Omega$. *induces the composition operator* C_T *if and only if for some* $k > 0$,

$$\mu T^{-1}(A) \leq k\mu(A).$$

for $A \in \mathcal{A}$.

Proof. Suppose that the composition operator is bounded on $L_{p,q;b}(\Omega, X)$. Then there exists $K > 0$ such that

$$\|C_T f\|_{p,q;b} \leq K\|f\|_{p,q;b}$$

Let x_0 be the fixed element of X with $\|x_0\| = 1$. Define the characteristic function χ_A for each measurable subset A of $\mathcal{A}$ by,

$$\chi_A(\omega) = \begin{cases} x_0, & \text{if } \omega \in A \\ 0, & \text{otherwise} \end{cases}$$

Then we find that

$$\|\chi_A\|^*(t) = \chi_{[0,\mu(A)]}(t)$$

and

$$\|\chi_A\|^{**}(t) = \begin{cases} 1, & \text{if } \mu(A) \le t \\ \dfrac{\mu(A)}{t}, & \text{if } \mu(A) \le t \end{cases}$$

This gives

$$\|\chi_A\|_{p,q;b} \approx (\mu(A))^{\frac{1}{p}} \left(\gamma_b^m(\mu(A))\right)$$

and

$$\|C_T\chi_A\|_{p,q;b} \approx \left(\mu\left(T^{-1}(A)\right)\right)^{\frac{1}{p}} \left(\gamma_b^m\left(\mu\left(T^{-1}(A)\right)\right)\right)$$

Thus, we get

$$\|C_T\chi_A\|_{p,q;b} \le K\|\chi_A\|_{p,q;b}$$

$$\left(\mu\left(T^{-1}(A)\right)\right)^{\frac{1}{p}} \left(\gamma_b^m\left(\mu\left(T^{-1}(A)\right)\right)\right)$$

$$\le K(\mu(A))^{\frac{1}{p}} \left(\gamma_b^m(\mu(A))\right)$$

$$\mu\left(T^{-1}(A)\right) \le k\mu(A), where$$

$$k = \left(\frac{\left(\gamma_b^m(\mu(A))\right)}{\left(\gamma_b^m\left(\mu\left(T^{-1}(A)\right)\right)\right)}\right)^p.$$

Conversely, suppose the given condition holds. Then

$$\mu_{C_T f}(s) = \mu\{\omega \in \Omega : \|C_T f\|(\omega) > s\}$$
$$= \mu\{\omega \in \Omega : \|f \circ T\|(\omega) > s\}$$
$$= \mu T^{-1}\{\omega \in \Omega : \|f\|(\omega) > s\}$$
$$\le k\mu\{\omega \in \Omega : \|f\|(\omega) > s\}$$
$$= \mu_f(s)$$

And

$$\|C_T f\|^*(t) = \inf\{s>0 : \mu_{C_T f}(s) \le t\}$$
$$\le \inf\{s>0 : k\mu_f(s) \le t\} = k\|f\|^*(t)$$

Thus

$$\|C_T f\|^*(t) \le k\|f\|^*(t).$$

Also

$$\|C_T f\|^{**} \le k\|f\|^{**}(t).$$

Therefore

$$\|C_T f\|_{p,q;b} = \left\| t^{\frac{1}{p}-\frac{1}{q}} \gamma_b(t) \|C_T f\|^{**}(t) L_q(0,\infty) \right\|$$

$$\le \left\| t^{\frac{1}{p}-\frac{1}{q}} \gamma_b(t) k \|f\|^{**}(t) L_q(0,\infty) \right\| = k\|f\|_{p,q;b}.$$

Thus, CT is a bounded operator on $L_{p,q;b}(\Omega, X)$.

Corollary 2.2. *A measurable transformation T induces the composition operator C_T on $L_{p,q;b}(\Omega, X)$ if and only if μT^{-1} is absolutely continuous with respect to μ and f_T belongs to $L^\infty(\Omega)$.*

Theorem 2.3. *If C_T is the composition operator on $L_{p,q;b}(\Omega, X)$. Then C_T is measure preserving if and only if C_T is an isometry.*

Proof. Suppose that T is measure preserving them

$$\mu T^{-1}(A) = \mu(A)$$

for all $A \in \mathcal{A}$

The distribution function of C_T becomes

$$\mu_{C_T f}(s) = \mu\{\omega \in \Omega : \|C_T f\|(\omega) > s\}$$
$$= \mu\{\omega \in \Omega : \|f \circ T\|(\omega) > s\}$$
$$= \mu\{\omega \in \Omega : \|f\|(\omega) > s\}$$
$$= \mu_f(s)$$

and

$$\|C_T f\|^*(t) = \|f\|^*(t)$$

Also

$$\|C_T f\|^{**} = \|f\|^{**}(t)$$

This gives

$$\|C_T f\|_{p,q;b} = \left\| t^{\frac{1}{p}-\frac{1}{q}} \gamma_b(t) \|C_T f\|^{**}(t) L_q(0,\infty) \right\|$$
$$= \left\| t^{\frac{1}{p}-\frac{1}{q}} \gamma_b(t) \|f\|^{**}(t) L_q(0,\infty) \right\|$$
$$= \|f\|_{p,q;b}.$$

Converse of the theorem is obvious.

Example 1. *Let* $\Omega = \mathbb{R}$. *with Lebesgue measure and X be any Banach space. Define*

$$T(\omega) = a\omega + b \quad a \neq 0,1$$

Then T is a non-singular transformation on Ω which is not measure preserving. Hence CT is not an isometry on $L_{p,q;b}(\Omega, X)$

Theorem 2.4. *If* C_T *is a composition operator on* $L_{p,q;b}(\Omega, X)$. *Then* C_T *has closed range if and only if there exists* $\epsilon > 0$ *such that* $f_T(\omega) \geq \epsilon$ *for almost all* $\omega \in S$, *the support of* f_T.

Proof. Suppose f_T is bounded away from zero then there exists a positive real number ϵ, such that

$$f_T(\omega) \geq \epsilon$$

for almost all $\omega \in S$.

$$\mu \circ T^{-1}(A) = \int_A f_T(\omega) d\mu \geq \epsilon\mu(A)$$

where

$$A = \{\omega \in S : |f(\omega)| > s\}$$

$$\mu_{C_T f}(s) = \mu\{\omega \in \Omega : |f \circ T(\omega)| > s\}$$

$$= \mu T^{-1}\{\omega \in S : |f(\omega)| > s\}$$

$$\geq \epsilon\mu\{\omega \in S : |f(\omega)| > s\}$$

$$\mu_{C_T f}(s) \geq \epsilon\, \mu_f(s)$$

And

$$\|f \circ T\|^*(\epsilon t) \geq \|f\|^*(t), \text{ for all } t \geq 0$$

And

$$\|f \circ T\|^{**}(\epsilon t) \geq \|f\|^{**}(t), \text{ for all } t \geq 0$$

This gives

$$\|C_T f\|_{p,q;b} \geq \epsilon^{\frac{1}{p}} \|f\|_{p,q;b}$$

Thus, CT has closed range.

Conversely, CT has closed range then there exists $\epsilon > 0$ such that

$$\|C_T f\|_{p,q;b} \geq \epsilon \|f\|_{p,q;b}$$

for all $f \in L_{p,q;b}(S)$

Choose a natural number n such that $\frac{1}{n} < \epsilon$. Let if possible, $\mu(B) > 0$ where $B = \left\{ x \in S : f_T(x) < \frac{1}{n^p} \right\}$.

Then $\mu(B) < \frac{1}{n^p} \mu(B)$ Then

$$\|\chi_B \circ T\|^*(t) \leq \frac{1}{n^p} \|\chi_B\|^*\left(n^p T\right), \textit{ for all } t > 0$$

and

$$\|\chi_B \circ T\|^{**}(t) \leq \frac{1}{n^p} \|\chi_B\|^{**}\left(n^p T\right), \textit{ for all } t > 0$$

This gives

$$\|C_T \chi_B\|_{p,q;b}^q \leq \frac{1}{n^p} \|\chi_B\|_{p,q;b}^q$$

which is a contradiction. Hence fT is bounded away from zero.

Theorem 2.5. *If* C_T *is a composition operator on* $L_{p,q;b}(\Omega, X)$. *Then* C_T *has dense range in* $L_{p,q;b}\left(\Omega, T^{-1}(\mathcal{A}), \mu\right)$.

Proof. We will consider two cases:

Case 1. When $\mu(\Omega) < \infty$.Then $\chi_A \in L_{p,q;b}\left(\Omega, T^{-1}(\mathcal{A}), \mu\right)$ so we can obtain $B \in \mathcal{A}$ such that

$$\chi_A = \chi_{T^{-1}(B)} = C_T \chi_B$$

Thus CT belong to the range of CT and hence all simple functions of $L_{p,q;b}\left(\Omega, T^{-1}(\mathcal{A}), \mu\right)$ belong to $R(C_T)$ where $R(C_T)$ denotes the range of CT. Hence, range of CT is dense in $L_{p,q;b}\left(\Omega, T^{-1}(\mathcal{A}), \mu\right)$.

Case 2. When $\mu(\Omega) = \infty$. Let $g \in R(C_T)$ then there is a sequence of functions $\langle g_n \rangle$ in $R(C_T)$ converging to g in $L_{p,q;b}(\Omega, \mathcal{A}, \mu)$. Since $g_n \in R(C_T)$,

$$g_n = C_T f_n$$

Clearly, each gn is $T^{-1}(\mathcal{A})$ measurable and hence g is also $T^{-1}(\mathcal{A})$ measurable. Now suppose $g = \chi_A$. By adjusting f on a set of measure zero, suppose $A = T^{-1}(B)$ for some $B \in \Omega$.Since $(\Omega, \mathcal{A}, \mu)$ is a σ-finite space

$$B = \bigcup_{n=1}^{\infty} B_n$$

where $\mu(B_n) < \infty$ for each n and $\langle B_n \rangle$ is an increasing sequence of measurable sets.

This gives

$$\left\| C_T \chi_{B_n} - f \right\| = \left\| C_T \chi_{B_n} - C_T \chi_B \right\| = \left\| \left(\chi_{B_n} - \chi_B \right) \circ T \right\|$$

which converges to zero. Thus R(CT) is dense in $L_{p,q;b}\left(\Omega, T^{-1}(\mathcal{A}), \mu\right)$.

Theorem 2.6. *T inducing the composition operator C_T on $L_{p,q;b}(\Omega, \mathcal{A}, \mu)$ is a surjection if and only if f_T is bounded away from zero on its support and $T^{-1}(\mathcal{A}) = \mathcal{A}$.*

Proof. Suppose C_T is a surjection. Then from the last theorem C_T has closed range if and only if f_T is bounded away from zero on its support. Let $A \in \mathcal{A}$ be of finite measure. Since C_T is a surjection there exist $f \in L_{p,q;b}(\Omega, \mathcal{A}, \mu)$ such that $C_T f = \chi_A$. Let

$$B = \{\omega : \omega \in \Omega \text{ and } f(\omega) = 1\}$$

Then

$$C_T \chi_B = \chi_A$$

Hence, $T^{-1}(B) = A$. Thus $\mathcal{A} \subseteq T^{-1}(\mathcal{A})$. Thus $\mathcal{A} = T^{-1}(\mathcal{A})$. Converse is obvious.

Corollary 2.7. *A composition operator C_T on B(Ω, X), has dense range if and only if $T^{-1}(\mathcal{A}) = \mathcal{A}$.*

Theorem 2.8. *If T induces a composition operator on ≐(Ω, X), then C_T^*, the adjoint of C_T is P_T*

Proof. Let *A e A* be such that $\mu(A) < \infty$. Then for $g \in L_{p',q';b^{-1}}$

$$\left(C_T^* Fg\right)(\chi_A) = F_g(C_T \chi_A) = \int (\chi_A \circ T).g d\mu$$
$$= \int E(g).\chi_A \circ T d\mu = \int E(g) \circ T^{-1}.\chi_A d\mu T^{-1}$$
$$= \int E(g) \circ T^{-1}.\chi_A f_T d\mu = \int F_{E(g) \circ T^{-1} f_T(\chi_A)}$$

By identifying $g \in L_{p',q';b^{-1}}$ with the functional $F_g \in \left(L_{p,q;b}\right)^* = L_{p',q';b^{-1}}$, we get

$$C_T^* g = \left(E(g) \circ T^{-1}\right).f_T = P_T g$$

Theorem 2.9. *If T induces a composition operator on B(Ω, X), then* $N\left(C_T^*\right)$ *is either zero dimensional or infinite dimensional.*

Proof. Suppose $g \in N\left(C_T^*\right)$ and $g \neq 0$. Let

$$A = \{\omega \in \Omega : g(\omega) \neq 0\}$$

then $\mu(A) \neq 0$. Let $\langle A_n \rangle$ be a sequence of disjoint measurable subsets of A such that

$$A = \bigcup_{n=1}^{\infty} A_n$$

where $\mu(A_n) < \infty$. For each $n \in \mathbb{N}$, let $g_n = g.\chi_A \circ T$. For each n,

$$C_T^*(g_n) f = \int (g.\chi_A \circ T)(f \circ T) d\mu$$
$$= \int g.(\chi_A f \circ T) d\mu = C_T^*(g)(\chi_A f) = 0.$$

Therefore $\{g_n : n \geq 1\}$ is a linearly independent subset of $N\left(C_T^*\right)$. Hence, if $N\left(C_T^*\right)$ is not zero dimensional, it is infinite dimensional.

Theorem 2.10. *If T induces a composition operator on B(Ω, X). Then* C_T *is invertible if and only if* C_T *is Fredholm.*

Proof. If C_T is invertible then C_T is Fredholm. Conversely, let C_T be Fredholm then $N(C_T)$ and $N\left(C_T^*\right)$ are both finite dimensional and are of zero dimension. Therefore C_T is injective and has dense range. Since $R(C_T)$ is closed, therefore C_T is surjective. Thus C_T is invertible.

Definition 2. *For a strongly measurable function* $f:\Omega\to B(X)$, *the set*

$$ess_f=\left\{\lambda\in\Omega:\mu\left(\left\|f(\omega)-\lambda\right\|<\epsilon\right)\neq 0\ \forall\ \epsilon>0\right\}$$

is called the essential range off.

Theorem 2.11. [11] *If* C_T *is a composition operator on* $L_{p,q;b}(\Omega,X)$, *then the following are equivalent:*

(i) C_T *is injective.*
(ii) *f and f °T have the same essential ranges for every* $f\in L_{p,q;b}(\Omega,X)$.
(iii) μ *is absolutely continuous with respect to* $\mu\circ T^{-1}$.
(iv) f_T *is different from zero almost everywhere.*

REFERENCES

1. S.C. Arora, G. Datt and S. Verma, Multiplication and composition operators on Lorentz-Bochner spaces, Osaka J. Math. 45 (2008), 629-641.
2. S.C Arora, G. Datt and S. Verma, Multiplication operators on Lorentz spaces, Indian J. Math. 48 (2006), 317-329.
3. P. Bala, A. Gupta, N. Bhatia, Multiplication operators on Orlicz-Lorentz sequence spaces, Int. Journal of Math. Analysis 7 (2013), no. 30, 1461-1469.
4. D.E. Edmunds and W.D. Evans, Hardy operators, function spaces and embeddings, Springer Monographs in Mathematics Springer-Verlag, Belin, 2004.
5. I. Eryilmaz, Multiplication operators on Lorentz-Karamata-Bochner spaces, Math. Slovaca 62 (2012), no. 2, 293-300.
6. B.S.Komal and S. Gupta, Multiplication operators between Orlicz spaces, Integral Equations Operator Theory 41 (2001), 324-330.
7. V.N. Mishra, Some problems on Approximations of Functions in Banach spaces, PhD Thesis (2007) Indian Instiiute of Technology, Roorkee-247 667, Uttarakhand, India.
8. V. N. Mishra, K. Khatri, L. N. Mishra, and Deepmala, Inverse result in simultaneous approximation by Baskakov-Durrmeyer- Stancu operators, Journal of Inequalities and Applications 2013, 2013: 586.
9. V. N. Mishra and L. N. Mishra, "Trigonometric Approximation of Signals (Functions) in $L_p\ (p\geq 1)$-Norm," International Journal of Contemporary Mathematical Sciences, Vol. 7, No. 19, 2012, pp. 909-918.
10. E.Seneta, Regularly Varying Functions, Lecture Notes in Math. 508, Springer-Verlag, Berlin, 1976.

11. R.K. Singh and J.S. Manhas. Composition operators on function spaces, North Math. Studies 179, North Holland, Amsterdam, 1993.

12. H. Takagi and K. Yokouchi, Multiplication and composition operators between two Lp-spaces, In: Function Spaces (Edwardsville, IL, 1998). Contemp. Math. 232, Amer. Math. Soc, Providence, RI, 1999, pp. 321-338.

9

COMMON FIXED POINT THEOREMS FOR TWO SELF-MAPS IN CONE METRIC SPACES UNDER DIFFERENT CONTRACTIVE CONDITIONS

P. Rama Bhadra Murthy*, M. Rangamma

* Department of Mathematics, University College of Science, Osmania University, Hyderabad, India

ABSTRACT

The existence of unique common fixed point theorems for two weakly compatible self-maps satisfying different contractive conditions in cone metric spaces without using normality.

INTRODUCTION

The fundamental work in metric fixed point theory, by Stefan Banach in 1922 is famous as Banach Contraction Principle and found many applications viz., in proving the existence and uniqueness of solutions of Differential, Integral, Integro-Differential, Impulsive differential equations, etc. A number of authors introduced different contractive type mappings and proved many fixed point theorems extending the theory. B.E.Rhoades [1], Paula Collaco and Jaime Carvalho E Silva [2] compared various definitions of contractive mappings. In 2007, Huang and Zhang [3] introduced the concept of cone metric spaces by replacing the codomain with Banach spaces in a metric function whose range satisfy the properties of a cone. Subsequently, Abbas and Jungck [4], Abbas and Rhoades [5] have studied common fixed point theorems in cone metric spaces for normal cones with the assumption of normality. Sh.Rezapour and R.Hamlbarani [6] proved some fixed point theorems for any cone, omitting the assumption of normality. Recently, several authors have proved and proving many common fixed point theorems. See [7-12].The purpose of this paper is to prove the common fixed point theorems for all cones which were proved by Abbas and Jungck in [4] only for normal cones.

The following definitions are taken from [3, 4, 5]. In the entire paper, E and R represents Banach space and Real Numbers respectively.

Definition 1.1: ([3]) Let E be a Banach space and $P \subseteq E$. P is called a cone if it satisfies the following properties

(i) P is closed, non-empty and $P \neq \{0\}$.
(ii) $a, b \in R, a, b \geq 0, x, y \in P$ *implies* $ax + by \in P$.
(iii) $x \in P$ *and* $-x \in P$ *implies* $x = 0$.

Examples 1.2:

(i) $E = R$ *and* $P = \{x \in E, x \geq 0\}$ is a cone.
(ii) Every non-zero Banach Space acts as a cone to itself.

Definition 1.3: ([3]) For a given cone P, define a partial ordering $\leq$ w.r.t. P by

(i) $x \leq y$ iff $y - x \in P$
(ii) $x < y$ *implies* $x \leq y$ *and* $x \neq y$
(iii) $x \ll y$ *implies* $y - x \in int\, P$ *where* $int\, P$ *denotes interior of* P.

Definition 1.4: ([3]) Let X be a non-empty set and E be a Banach Space. Suppose the mapping $d : XxX \to E$ satisfy

(i) $0 \leq d(x, y) \forall x, y \in X$ *i.e.,* $d(x, y) \in P$

(ii) $d(x,y) = 0$ *iff* $x = y$

(iii) $d(x, y) = d(y, x) \forall x, y \in X$

(iv) $d(x, y) \leq d(x, z) + d(z, y) \; \forall x, y, z \in X$.

Then d is called a cone metric on X and (X, d) is called a cone metric space.

Example 1.5: $E = X = R, P = \{x \in E : x \geq 0\}$, and $d : XxX \to E$ defined by $d(x, y) = |x - y|$.

Then (X, d) is a cone metric space.

Definition 1.6: ([3]) Let (X, d) be a cone metric space, $x \in X$ and $\{x_n\}_{n \geq 1}$ be a sequence in X. Then

(i) $x_n \to x$ whenever for every $c \in E$ with $0 \ll c, \exists$ a natural number N such that $d(x_n, x) \ll c \forall n \geq N$.
(ii) $\{x_n\}$ is a Cauchy Sequence if for every $c \in E$ with $0 \ll c$ there is a natural number s N such that $d(x_n, x_m) \ll c \forall n, m \geq N$.
(iii) (X, d) is a complete cone metric space if every Cauchy sequence in X is convergent in X.

The following lemma's are taken from [3]

Lemma 1.7: ([3]) Let (X,d) be a cone metric space. Let P be a normal cone with normal constant K. Let $\{x_n\}$ be a sequence in X Then

(i) $x_n \to x$ *iff* $d(x_n,x) \to 0$ *as* $n \to \infty$.

(ii) $\{x_n\}$ is a Cauchy sequence iff $d(x_n,x_m) \to 0$ as $n,m \to \infty$.

Lemma 1.8: ([3]) Let (X,d) be a cone metric space. Let $\{x_n\}$ be a sequence in X If $\{x_n\}$ converges to x and $\{x_n\}$ converges to y, then $x = y$ i.e., the limit of the sequence is unique.

For the definition and examples of normal and non-normal cones, see [3, 4, 5].

Definition 1.9: ([4]) Let X be a non-empty set and f, g two self maps on X. Then

(i) If $q = fp = gp$ for some $p \in X$, then p is called a coincidence point of f and g. Also q is called a point of coincidence of f and g.

(ii) If $p = fp = gp$ for some $p \in X$, then p is called a common fixed point of f and g.

Definition 1.10: ([4]) Let X be a non-empty set and f, g two self maps on X The pair $\{f, g\}$ is said to be weakly compatible if $f(gt) = g(ft)$ whenever $ft = gt$ for some $t \in X$.

Lemma 1.11: ([4]) Let f and g be weakly compatible self maps of a set X If f and g have a unique point of coincidence $w = fx = gx$, then w is the unique common fixed point of f and g.

MAIN RESULTS

Theorem 2.1: Let (X,d) be a cone metric space. Suppose the self-maps $f, g : X \to X$ satisfy the contractive condition

$$d(fx, fy) \leq kd(gx, gy) \;\; \forall \, x, y \in X \tag{2.1)(1}$$

where $k \in [0,1)$ is a constant. If the range of f is contained in range of g and range of g is a complete subspace of X More over if f and g are weakly compatible, f and g have a unique common fixed point.

Proof: Let $x_0 \in X$ be arbitrary. Choose $x_1 \in X$ such that $f(x_0) = g(x_1)$. This possible since $f(X) \subseteq g(X)$..Continuing this process, choose $x_{n+1} \in X$ such that

$$f(x_n) = g(x_{n+1}) \tag{2.1)(2}$$

Now,

$$d(gx_{n+1}, gx_n) = d(fx_n, fx_{n+1}) \leq kd(gx_n, gx_{n-1})$$

using 2.1(1)

By Repeated application of 2.1(1), we get

$$d(gx_{n+1}, gx_n) \leq k^n d(gx_1, gx_0) \tag{2.1)(3}$$

For $n > m$,

$$\begin{aligned} d(gx_n, gx_m) &\leq d(gx_n, gx_{n-1}) + d(gx_{n-1}, gx_{n-2}) \\ &\quad + \ldots + d(gx_{m+1}, gx_m) \\ &\leq (k^{n-1} + k^{n-2} + \ldots + k^m)\, d(gx_1, gx_0) \\ &\leq k^n \left(1 + k + k^2 + \ldots + \ldots\right) d(gx_1, gx_0) \\ &= \frac{k^m}{1-k} d(gx_1, gx_0) \end{aligned} \tag{2.1)(4}$$

Let $0 \ll c$ be given. Choose $n_1, \in N$ such that

$$\frac{k^m}{1-k} \; d(gx_1, gx_0) \ll c \forall m \geq n_1 \in N \tag{2.1)(5}$$

From 2.1(4) and 2.1(5), we get

$$d(gx_n, gx_m) \ll c \forall m, n \geq n_1 \in N$$

$\Rightarrow \{gx_n\}$ is a Cauchy sequence in $g(X)$, be def. of 1.6(ii)

$\Rightarrow \{gx_n\}$ is a convergent sequence in $g(X)$ since $g(X)$ is complete in X.

Let $\{gx_n\}$ is convergent to $q \in g(X)$. Consquently, there is

$$p \in X \;\text{ such that } gp = q \tag{2.1)(6}$$

For the same given $c \in E$, choose $n_2 \in N$ such that

$$d(gx_{n-1}, gp) \ll \frac{c}{k} \ \forall\, n \geq n_2 \in N \qquad 2.1(7) \qquad (2.1)(7)$$

Hence, using 2.1(7)

$$d(gx_n, fp) = d(fx_{n-1}, fp) \leq kd(gx_{n-1}, gp) \ll c,$$
$$\Rightarrow d(gx_n, fp) \ll c \ \forall\ n \geq n_2 \in N$$
$$\Rightarrow gx_n \to fp.$$

Hence $\{gx_n\}$ converges to both q and fp. By uniqueness property of limit

$$fp = gp = q \qquad (2.1)(8)$$

$\Rightarrow q$ is the point of coincidence of f and g

Let $r \in X$ be any other coincidence point of f and g

$\Rightarrow fr = gr$ be the point of coincidence of f and g 2.1(9)

Now, $d(gr, gp) = d(fr, fp) \leq kd(gr, gp)$

$\Rightarrow d(gr, gp) \leq kd(gr, gp)$

$\Rightarrow (k-1)d(gr, gp) \in P$ using 1.3(i)

Multiplying with positive real number $(1-k)$, we get $-d(gr, gp) \in P$ But, we have $d(gr, gp) \in P$..

From the definition of cone and cone metric, we get

$$gr = gp. \qquad (2.1)(10)$$

From 2.1(8, 9, 10), f and g have unique point of coincidence.

Finally, let f and g are weakly compatible self-maps having unique point of coincidence. Using the lemma 1.11,

f and g have a unique common fixed point.

Example 2.1.1: Let $X = E = R, P = \{x \in E : x \geq 0\}$,

$d : RxR \to E$ such that $d(x, y) = |x - y|$. Define

$$fx = \frac{\alpha}{\beta+1} x \ \text{if}\ x \neq 0 \ \text{and}\ fx = \gamma \ \text{if}\ x = 0,,$$
$$gx = \alpha x \ \text{if}\ x \neq 0 \ \text{and}\ gx = \gamma \ \text{if}\ x = 0,$$

where α *is constant* $\beta \geq 1$ *and* $\gamma \neq 0$.

Clearly $d(fx, fy) \leq kd(gx, gy) \; \forall \; x, y \in X$

where : $k = \frac{1}{\beta} \in (0,1]$ is a cone metric space.

Moreover, $0 \in X$ is coincidence point for f and g But f and g are not weakly compatible at $0 \in X$. Hence f and g do not have common fixed point.

Theorem 2.2: Let (X, d) be a cone metric space. Suppose the maps $f, g : X \to X$ satisfy the contractive condition

$$d(fx, fy) \leq k\left[d(fx, gx) + d(fy, gy)\right] \; \forall x, y \in X \quad (2.2)(1)$$

where $k \in [0, \frac{1}{2})$ is a constant. If the range of f is contained in range of g and the range of g is a complete subspace of X, then f and g have a unique coincidence point in X. Moreover, if f and g are weakly compatible, $f$and g have a unique common fixed point.

Proof: Let $x_0 \in X$ be arbitrary. Choose $x_1 \in X$ such that $f(x_0) = g(x_1)$ whitch is possible since $f(X) \subseteq g(X)$. Continuing this process, choose $x_{n+1} \in X$ such that

$$f(x_n) = g(x_{n+1}) \quad (2.2)(2)$$

Now,

$$\begin{aligned}
d(g_{n+1}, gx_n &= d \; fx_n \; fx_{n-1} \\
&\leq k[d(fx_n, gx_n) + d(fx_{n-1}, gx_{n-1})] \\
&\leq k[d(gx_{n+1}, gx_n) + d(gx_n, gx_{n-1})] \\
\Rightarrow d(gx_{n+1}, gx_n) &\leq \frac{k}{1-k} d(gx_n, gx_{n-1})
\end{aligned}$$

$= hd(gx_n, gx_{n-1})$; where $h = \frac{k}{1-k}$

By repeated application of 2.2(1), we get

$$d(gx_{n+1}, gx_n) \leq h^n \; d(gx_1, gx_0) \quad (2.2)(3)$$

For $n > m$,

$$
\begin{aligned}
d(gx_n, gx_m) &\le d(gx_n, gx_{n-1}) + d(gx_{n-1}, gx_{n-2}) \\
&\quad + \ldots + d(gx_{m+1}, gx_m) \\
&\le \left(h^{n-1} + h^{n-2} + .. + h^m\right) d(gx_1, gx_0) \\
&\le h^m \left(1 + h + h^2 + ... + \ldots\right) d(gx_1, gx_0) \\
&= \frac{h^m}{1-h}\, d(gx_1, gx_0), \text{ since } |h| < 1
\end{aligned}
\tag{2.2)(4}
$$

Let $0 \ll c$ be given. Choose $n_1 \in N$ such that

$$\frac{h^m}{1-h}\, d\left(gx_1, gx_0\right) \ll c \; \forall\; m \ge n_1 \in N \tag{2.2)(5}$$

From 2.2(4), 2.2(5), we get

$$d\left(gx_n, gx_m\right) \ll c \forall m, n \ge n_1 \in N$$

$\Rightarrow \{gx_n\}$ is a Cauchy sequence in $g(X)$.

$\Rightarrow \{gx_n\}$ is a convergent sequence in $g(X)$ since $g(X)$ is complete in X.

Let $\{gx_n\}$ is convergent to $q \in g(X)$ Consequently there is $p \in X$ such that

$$gp = q \tag{2.2)(6}$$

For the same given $c \in E$, choose $n_2 \in N$ such that

$$d\left(gx_n, gx_{n-1}\right) \ll \frac{(1-k)c}{2k}$$

and

$$d\left(gx_n, gp\right) \ll \frac{(1-k)c}{2} \forall n \ge n_2 \in N \tag{2.2)(7}$$

Hence,

$$d(gx_n, fp) = d(fn_{n-1}, fp)$$
$$\leq k[d(fx_{n-1}, gx_{n-1}) + d(fp, gp)]$$
$$\leq k[d(gx_n, gx_{n-1}) + d(fp, gx_n) + d(gx_n, gp)]$$
$$\Rightarrow d(gx_n, fp) \leq \frac{1}{1-k}[kd(gx_n, gx_{n-1}) + d(gx_n, gp)]$$

Applying 2.2(7), we get

$$d(gx_n, fp) \ll c \ \forall \ n \geq n_2 \in N$$
$$\Rightarrow gx_n \to fp \tag{2.2)(8}$$

$\Rightarrow \{gx_n\}$ converges to both q and fp.

By uniqueness property of limit,

$$fp = gp = q. \tag{2.2)(9}$$

$\Rightarrow fp = gp$ is point of coincidence of f and g.

Let $r \in X$ be any other coincidence point of f and g

$\Rightarrow fr = gr$ is the point of coincidence of f and g

Now

$$d(gr, gp) = d(fr, fp)$$
$$\leq k\left[d(fr, gr) + d(fp, gp)\right] = 0$$
$$\Rightarrow d(gr, gp) \leq 0$$
$$\Rightarrow -d(gr, gp) \in P.$$

We have $d(gr, gp) \in P$.

From the definition of cone and cone metric, we get

$$gr = gp. \tag{2.2)(10}$$

From 2.2(8, 9, 10), f and g have unique point of coincidence.

Finally, let f and g are weakly compatible self-maps having unique point of coincidence. Using the lemma 1.11,

f and g have a unique common fixed point.

Example 2.2.1: Since $|x-y| \le |x|+|y| \; \forall \; x, y \in R$, example 2.1.1 also satisfies the contractive condition 2.2(1) with the same $k = \frac{1}{\beta} \in (0,1]$.

Theorem 2.3: Let (X,d) be a cone metric space. Suppose the mappings $f, g: X \to X$ satisfy the contra

$$d(fx, fy) \le k\left[d(fx, gy) + d(fy, gx)\right] \; \forall \; x, y \in X \tag{2.3)(1}$$

where $k \in [0, \frac{1}{2})$ s constant. If tie range of f is contained in range of g and $g(X)$ is a complete subspace of X , then f and g have a unique coincidence point in X. Moreover, if f and g are weakly compatible, f and g have a unique common fixed point.

Proof: Let $x_0 \in X$ be arbitrary. Choose $x_1 \in X$ such that $f(x_0) = g(x_1)$ which is possible since $f(X) \subseteq g(X)$.

Continuing this process, choose $x_{n+1} \in X$ such that

$$f(x_n) = g(x_{n+1}) \tag{2.3)(2}$$

Now,

$$d(gx_{n+1}, gx_n) = d(fx_n, fx_{n-1})$$
$$\le (fx_n, fx_{n-1}) + d(fx_{n-1}, gx_n)]$$
$$\le k(d(gx_{n+1}, gx_{n-1}) + d(gx_n, gx_n)]$$
$$\le k\left[d(gx_{n+1}), gx_n) + d(gx_n, gx_{n-1})\right]$$
$$\Rightarrow d(gx_{n+1}, gx_n) \le \frac{k}{1-k} d(gx_n, gx_{n-1})$$

$= hd(gx_n, gx_{n-1})$. where $h = \frac{k}{1-k}$.

By repeated application of 2.2(1), we get

$$d(gx_{n+1}, gx_n) \le h^n \, d(gx_1, gx_0) \tag{2.3)(3}$$

For $n > m$,

$$\begin{aligned} d(gx_n, gx_m) &\le d(gx_n, gx_{n-1}) + d(gx_{n-1}, gx_{n-2}) \\ &\quad + \ldots + d(gx_{m+1}, gx_m) \\ &\le \left(h^{n-1} + h^{n-2} + .. + h^m\right) d(gx_1, gx_0) \qquad (2.3)(4) \\ &\le h^m \left(1 + h + h^2 + \ldots + \ldots\right) d(gx_1, gx_0) \end{aligned}$$

$= \dfrac{h^m}{1-h} d(gx_1, gx_0)$ since $|h| < 1$

Let $0 \ll c$ be given. choose $n_1 \in N$ such that

$$\frac{h^m}{1-h} d\left(gx_1, gx_0\right) \ll c \;\forall\; m \ge n_1 \in N. \qquad (2.3)(5)$$

From 2.3(4), 2.3(5), we get

$$d\left(gx_n, gx_m\right) \ll c \;\forall\; m, n \ge n_1 \in N$$

$\Rightarrow \{gx_n\}$ is a Cauchy sequence in $g(X)$

$\Rightarrow \{gx_n\}$ is a convergent sequence in $g(X)$ since $g(X)$ is complete in X.

Let $\{gx_n\}$ is convergent to $q \in g(X)$. Consequently there is $p \in X$ such that

$$gp = q \qquad (2.3)(6)$$

For the same given $c \in E$, choose $n_2 \in N$ such that

$$d\left(gx_n, gx_{n-1}\right) \ll \frac{c(1-k)}{2k}$$

And

$$d\left(gx_n, gp\right) \ll \frac{c(1-k)}{2k} \;\forall\; n \ge n_2 \in N \qquad (2.3)(7)$$

Hence,

$$d(gx_n, fp) = d(fx_{n-1}, fp)$$
$$\leq k\left[d(fx_{n-1}, gp) + d(fp, gx_{n-1})\right]$$
$$\leq k\left[d(gx_n, gp) + d(fp, gx_{n-1})\right]$$
$$\leq k\left[d(gx_n, gp) + d(fp, gx_n) + d(gx_n, gx_{n-1})\right]$$
$$\Rightarrow d(gx_n, fp) \leq \frac{k}{1-k}[d(gx_n, gp) + d(gx_n, gx_{n-1})]$$

$$\ll \frac{c}{2} + \frac{c}{2} = c \; \forall \; n \geq n_2 \in N \text{ from 2.3(7)}$$

$$\Rightarrow gx_n \rightarrow fp \qquad (2.3)(8)$$

$\Rightarrow \{gx_n\}$ converges to both q and fp.

By uniqueness property of limit,

$$fp = gp = q \qquad (2.3)(9)$$

$\Rightarrow fp = gp$ is point of coincidence of f and g.

Let $r \in X$ be any other coincidence point of f and g

$\Rightarrow fr = gr$ is the point of coincidence of f and g.

Now

$$d(gr, gp) = d(fr, fp) \leq k\left[d(fr, gp) + d(fp, gr)\right]$$
$$\leq k[d(gr, gp) + d(gp, gr)]$$
$$= 2kd(gr, gp)$$
$$\Rightarrow d(gr, gp) \leq 2kd(gr, gp)$$
$$\Rightarrow (2k-1)d(gr, gp) \in P$$

Multiplying with positive real number $(1-2k)$ we get $-d(gr, gp) \in P$ But we have $d(gr, gp) \in P$.

From the definition of cone and cone metric, we get

$$gr = gp. \qquad (2.3)(10)$$

From 2.2(8, 9, 10), f and g have unique point of coincidence.

Finally, let f and g are weakly compatible self-maps having unique point of coincidence. Using the lemma 1.11,

f and g have a unique common fixed point.

Example 2.3.1: Let $X = E = R, P = \{x \in E : x \geq 0\}$,

$d : RxR \to E$ such that $d(x, y) = |x - y|$.

Define $fx = \frac{x}{2}$ if $x \neq 0$ and $fx = \gamma$ if $x = 0$ and

$gx = x$ if $x \neq 0$ and $gx = \gamma$ if $x = 0$

Clearly f and g satisfies the contractive condition 2.3(1) with $k = \frac{1}{3} \in (0,1]$ Moreover, $0 \in X$ is coincidence point for f and g. But f and g are not weakly compatible at $0 \in X$ Hence f and g do not have common fixed point.

ACKNOWLEDGEMENT

The author thanks the assistance provided by the CSIR in all the respects.

REFERENCES

1. B.E.Rhoades, "A comparision of various definitions of contractive mappings", Trans. Amer. Math. Soc.26(1977) 257-290.
2. Paula Collaco and Jaime Carvalho E Silva,"A complete comparision of 25 contraction conditions", Nonlinear Analysis, Theory, Methods and Applications, Vol.30, No.1, pp.471-476.
3. L.G.Huang, X.Zhang, "Cone metric spaces and fixed point theorems of contractive mappings", J. Math. Anal. Appl.332(2) (2007) 1468-1476.
4. M.Abbas and G. Jungck, "Common fixed point results for non-commutating mappings withour continuity in cone metric spaces", J. Math. Anal. Appl. 341(2008)416-420.
5. M.Abbas and B.E.Rhoades, "Fixed and Periodic point results in Cone metric spaces", Appl. Math. Lett. 22(2009), 511-515.
6. S.Rezapour and Halmbarani, "Some notes on the paper "cone metric spaces and fixed point theorems of contractive mappings" ,J. Math. Anal. Appl. 345(2008)719-729.
7. S.Rezapour, R.H.Hagi, "Fixed points of multi functions on cone metric spaces", Number. Funct. Anal. 30(7-9) (2009)825-832.
8. Stojan Radenovic, "Common fixed points under contractive conditions in cone metric spaces", Computers and Mathematics with Applications. 58(2009)1273-1278.

9. Vetro, "Common fixed points in cone meric spaces", Rend. Ciric. Mat. Palermo. 56(2007)464-468.

10. T. Abdeljawad, E. Karapinar and K. Tas, "Common fixed point theorems in cone Banach spaces", Hacet. J. Math. Stat. 40(2011), 211-217.

11. Jankovic, S.Dadelburg, Z.Radenovic, "On cone metric spaces, A survey", Nonlinear Analysis, 74(2011), 2591-2601.

12. P. Raja, S.M. Vaezpour, "Some extensions of Banach's contraction principle in complete cone metric spaces", Fixed Point Theory and Applications, Article ID768294(2008), 11pages.

10

ON THE ANALYTIC CURVE OF C^2 WHICH IS NOT OMITTED BY EVERY FATOU-BIEBERBACH DOMAIN

Yukinobu Adachi[1]

[1] Yukinobu Adachi, Kurakuen, Nishinomiya, Hyogo, Japan

ABSTRACT

Let *C* be an irreducible (may be transcendental) analytic curve whose genus is geater than 1. Then every Fatou-Bieberbach domain does not omit *C*.

INTRODUCTION

We call that a domain $\Omega \subsetneq \mathbf{C}^2$ is a Fatou-Bieberbach domain if Ω is biholomorphic to C^2.

We call that an irreducible algebraic curve A of C^2 is hyperbolic when $g(A)>0$ or else $g(A)=0$ and the set $A \cap L_\infty$ consists more than two different points, where $g(A)$ is the genus of A and L_∞ is the line at infinity.

In the previous paper [2], we proved the case A (theorem 2.1). In this paper, every Fatou-Bieberbach domain does not omit C which is the same in Abstract (Theorem 3.12). Moreover we prove that the property that an analytic curve of C2 does not omit every Fatou-Bieberbach domain is unchanged one by a transformation of Aut(C2) (Theorem 3.7). Therefore every Fatou-Bieberbach domain does not omit T(A) with T $\in$ Aut(C2). And we prove also that there is an algebraic type curve of C2 which is isomorphic to an algebraic curve of any topological type such that it can not be transformed to an algebraic curve by any T $\in$ Aut(C2) (Proposition 3.3).

PRELIMINARY

Theorem 2.1 (Theorem 2.1 in [2]). Let A be a hyperbolic algebraic curve of C2. Then every Fatou-Bieberbach domain Ω does not omit A.

Remark 2.2 (Theorem 2.2 in [2]). Let Ω0 be every Fatou-Bieberbach domain in C2 arising from a polynomial automorphism, namely a polynomial basin. Then Ω0 does not omit every algebraic curve of C2.

Theorem 2.3 (Theorem 4.1 in Buzzard and Fornaess [4]). Let X be an arbitrary analytic curve of C2. Then there is a Fatou-Bieberbach domain Ω with $\Omega \supset X$ and $\Omega \neq C^2$ where some $\Phi : \Omega \to C^2$ is biholomorphic such that every nonconstant holomorphic map $f : C \to C^2$ intersects with Φ(X) at in_nite points and C2 − Φ(X) is Kobayashi hyperbolic.

Example 2.4 (Example 9.7 in [6]). There is a Fatou-Bieberbach domain Ω0 such that $\bar{\Omega}_0$ namely the closure of Ω_0 omits an algebraic complex line L.

Proposition 2.5 (Proposition 3.1 in [2]). For some transendental complex line $\tilde{L}$ such that there is a Fatou-Bieberbach domain Ω0 with $\bar{\Omega}_0 \cap \tilde{L} = \emptyset$.

From Theorem 2.3, Example 2.4 and Proposition 2.5, following theorem is easy to see.

Theorem 2.6. Let C be an analytic curve of C2. Then every Fatou-Bieberbach domain does not omit C or some (not all) Fatou-Bieberbach do-main Ω0 omits C.

CONCLUSION

Proposition 3.1. Let $\tilde{A}$ be a transendental hyperbolic curve of C2 which is isomorphic to a hyperbolic algebraic curve A. If $\tilde{A}$ is transformed to an algebraic hyperbolic curve A of C2 by T ∈ Aut(C2), then every Fatou-Bieberbach domain does not omit $\tilde{A}$.

Proof. We assume that there is a Fatou-Bieberbach domain Ω with $\Omega \cap \tilde{A} = \emptyset$. Then $T(\Omega) \cap A = \emptyset$. and T(Ω) is biholomorphic to C2 by T ∘ Φ, where Φ is a biholomorphic map of Ω to C2. Therefore T(Ω) is a Fatou-Bieberbach domain and it contradicts to Theorem 2.1.

Proposition 3.2. Let X and Ω be the same of Theorem 2.3. Then Φ(X) can not be transformed to any algebraic curve A which is hyperbolic or non by any T ∈ Aut(C2).

Proof. We assume that Φ(X) is transformed to an algebraic curve A by some T ∈ Aut(C2). As T(Ω) is a Fatou-bieberbach domain with $T \circ \Phi(X) = A$. Let $f: \mathbf{C} \to \mathbf{C}^2$ be a map to an algebraic complex line. Then $(T \circ \Phi(X)) \cap f(\mathbf{C})$ is a finite set of points at most and $\Phi(X) \cap (T{-}1 \circ f(\mathbf{C}))$ is also. It is a contradiction to Theorem 2.3.

Proposition 3.3 (cf. Proposition 3.12 in [2] . There is a transcendental analytic curve of C2 which is isomorphic to an algebraic curve of any topological type, that is an algebraic type curve of any type, such that it can not be transformed to an algebraic curve by any T ∈ Aut(C2). If we take X an algebraic type curve of any topological type, Φ(X) which is the same notation of Theorem 2.3 is such one.

Proposition 3.4. Let X, Ω and Φ be the same of Theorem 2.3. Then every Fatou-Bieberbach domain Ω′ does not omit Φ(X).

Proof. If some Ω′ omits $\Phi(X), \Phi'^{-1}: C^2 \rightarrow \Omega' \subset C2-\Phi(X)$ where Φ′ is a biholomorphic map of Ω′ to C2. Since C2−Φ(X) is Kobayashi hyperbolic, Φ′−1 is a constant map. It is a contradiction.

Proposition 3.5. Let C be an analytic curve of C2 which is transformed by T ∈ Aut(C2) to some analytic curve C′of C2 which does not omit every Fatou-Bieberbach domain. Then every Fatou-Bieberbach domain does not omit C.

Proof. We assume that there is a Fatou-Bieberbach domain Ω0 with $\Omega_0 \cap C = \emptyset$ Then $T(\Omega_0) \cap T(C) = \emptyset$. Since T(Ω0) is a Fatou-Bieberbach domain which omits C′ = T(C). It contradicts to the property of C.

Corollary 3.6. Let C be an analytic curve of C2 which does not omit every Fatou-Bieberbach domain. Then T(C) with every T ∈ Aut(C2) does not omits every Fatou-Bieberbach domain.

Proof. Since T−1 ∘ T(C) = C, T(C) does not omit every Fatou-Bieberbach domain also by Proposition 3.5.

From Proposition 3.5 and Corollary 3.6 it is easy to see the following theorem.

Theorem 3.7. The property that the analytic curve of C2 does not omit every Fatou-Bieberbach domain is unchanged one by a transformation of Aut(C2).

From Proposition 3.4 and Theorem 3.7 it is easy to see the following corollary.

Corollary 3.8. Let C be an analytic curve of C2 which is transformed by some T ∈ Aut(C2) to Φ(X) where X, Ω and Φ are the same of Proposition 3.4. Then every Fatou-Bieberbach domain Ω′ does not omit C.

Problem 3.9. Is there $\tilde{A}$ which is not transformed to an algebraic hyperbolic curve A of C2 or Φ(X), which is the same notation of Theorem 2.3, by some T ∈ Aut(C2)?

Proposition 3.10. Let C be an irreducible analytic curve of C2 with g(C) > 1. Then D = C2 − C is not biholomorphic to C2, that is, D is not a Fatou-Bieberbach domain.

Proof. We assume that there is a biholomorphic map Φ of D to C2(x, y) such as Φ : $x = \xi(z,w),\ y = \eta(z,w)$ where (z,w) ∈ D. Let P(x, y) be a nonconstant primitive polynomial, that is all level curve of {P(x, y) = α} is irreducible except at most finite number of $\alpha 1, \cdots, \alpha n$ .Then $P(\xi(z,w), \eta(z,w)) \in \mathcal{O}(D)$ is an essential singular curve of P(ξ, η), every level curve of P(ξ, η) can not be analytically continued to every point of C except at most one value α0 by well known Thullen's theorem.

Since $\{P(\xi, \eta) = \alpha, \alpha \neq \alpha_0, \alpha_1, \cdots, \alpha_n\}$ is considered as a Riemann surface R0 = R − {p1, · · · , pm} where R is a compact Riemann surface, that is an algebraic type Riemann surface such as π : R0 → {P(ξ, η) = α} ⊂C2 is the normalization. And at some pi ∈ {p1, · · · , pm} the cluster set of π in C2 is C. Because it is a pseudoconcave set of C2 by Theorem 3.4 in [1] and it is a degeneration locus of Kobayashi pseudodistance by Theorem 3.6 in [1]. It is a contradiction to g(C) > 1 by Theorem 4 in [3] .

Therefore C is not an essential singular curve of P(ξ, η) , that is, P(ξ, η) is at most meromorphically continued to C. Since P(x, y) ∈ $\mathcal{O}$(C2), P(ξ, η) is holomorphically continued to C. We set such function as F(z,w).

Since every level curve of F(z,w) is holomorphically isomorphic to algebraic type Riemann surface and an analytic cuve of C2, F(z,w) is an algebraic type entire function of Nishino's sence [5] , namely in the class (A). Then by principal theorem of [5] , F = φ ∘ Q ∘ T where φ is a polynomial function of one complex variable because P(x, y) is primitive, Q is a primitive polynomial and T ∈ Aut(C2).

Then T|D = Φ. It contradicts to the assumption. Then D is not biholomorphic to C2.

Proposition 3.11. Let C be an irreducible analytic curve of C2 such that D = C2 − C is not a Fatou-Bieberbach domain. Then every subdomain of D′ of D is not a Fatou-Bieberbach domain.

Proof. If D′ is a Fatou-Bieberbach domain, there is a biholomorphic map Φ of C2 to D′. Let I be an inclusion map of D′ to D and {f = 0} = C where f ∈ $\mathcal{O}$(C2). Since g = f ∘ I ∘ Φ is an entire function of C2 and g ≠ 0. There is a complex line L of C2 and g|L is considered as a transendental entire function of C, it is a contradiction by little Picard theorem because I ∘ Φ|L is an one to one map.

From Proposition 3.10 and 11, following theorem is easy to see.

Theorem 3.12. Let C be the same of Proposition 3.10.

Then every Fatou-Bieberbach domain does not omit C.

REFERENCES

1. Y. Adachi, A generalization of the big Picard theorem, Kodai Math. J., 18(1995), 408-424.
2. Y. Adachi, Remarks about Fatou-Bieberbach domains and algebraic or non-algebraic curves in C2, Far East J. Math. Soc. (FJMS), 34 (2009), 369-376.
3. Y. Adachi and Masakazu Suzuki, Degeneracy points of the Kobayashi pseudodistances on complex manifolds, Proceedings of Symposia in Pure Math., 52 (1991), Part 2, 41-51.
4. G. T. Buzzard and J. E. Fornaess, An embedding of C in C2 with hyperbolic complement, Math. Ann. 306 (1996), 539-546.
5. T. Nishino, Nouvelles recherches sur les fonctions entières de plusieurs variables conplexes (V) Fonctions qui se réduisent aux polynômes, J. Math. Kyoto Univ., 15 (1975), 527-553.
6. J. P. Rosay and W. Rudin, Holomorphic maps from Cn to Cn, Trans. A. M. S., 310 (1988), 47-86.

11

INEQUALITIES FOR THE STH DERIVATIVE OF POLYNOMIALS NOT VANISHING INSIDE A CIRCLE

GULSHAN SINGH

[1] Govt. Department of Education Jammu and Kashmir, India

ABSTRACT

Let P(z) be a polynomial of degree n having all its zeros in $|z| \geq k, k \geq 1$, then for $1 \leq R \leq k$, Bidkham and Dewan [J. Math. Anal. Appl. 166(1992), 191-193] proved max

$$\max_{|z|=R}|P'(z)| \geq \frac{n(R+k)^{n-1}}{(1+k)^n}\max_{|z|=1}|P(z)|$$

In this paper, we prove an interesting generalization as well as an improvement of this result by considering the sth derivative of lacunary type of polynomials P(z) of degree n > 3.

INTRODUCTION AND STATEMENT OF RESULTS

Let $P(z) := \sum_{j=0}^{n} a_j z^j$ be a polynomial of degree n and P'(z) its derivative, then it is known that

$$\max_{|z|=1}|P'(z)| \leq \max_{|z|=1}|P(z)|. \tag{1}$$

The above result, which is an immediate consequence of Bernstein's inequality on the derivative of a trigonometric polynomial is best possible with equality holding for the polynomial $P(z)=\lambda z^{n}$, where λ is a complex number.

If we restrict ourselves to the class of polynomials having all their zeros in $|z|\geq 1$, inequality (1) can be sharpened. In fact, Erdös conjectured and later Lax [6] proved that if P(z)≠ 0 in $|z|<1$, then

$$\max_{|z|=1}|P'(z)| \leq \frac{n}{2}\max_{|z|=1}|P(z)|. \tag{2}$$

On the other hand, if the polynomial P(z) of degree n has all its zeros in $|z|\leq 1$, then it was proved by Turán[10], that max

$$\max_{|z|=1}|P'(z)| \geq \frac{n}{2}\max_{|z|=1}|P(z)|. \tag{3}$$

Inequalities (2) and (3) are best possible and become equality for polynomials which have all zeros on $|z|=1$.

Inequality (2) was refined by Aziz and Dawood [1] by showing that under the same hypothesis that

$$\max_{|z|=1}|P'(z)| \leq \frac{n}{2}\left\{\max_{|z|=1}|P(z)| - \min_{|z|=1}|P(z)|\right\}. \tag{4}$$

Equality in (4) holds for $P(z)=\alpha+\beta z^{n}, |\alpha|\geq|\beta|$.

For the class of polynomials P(z) of degree n having all their zeros in $|z|\geq k, k\geq 1$, Malik [7] proved:

$$\max_{|z|=1}|P'(z)| \leq \frac{n}{1+k}\max_{|z|=1}|P(z)|. \tag{5}$$

Inequality (5) was further improved by Govil [5] who under the same hypothesis proved:

$$\max_{|z|=1}|P'(z)| \leq \frac{n}{1+k}\left\{\max_{|z|=1}|P(z)| - \min_{|z|=k}|P(z)|\right\}. \tag{6}$$

Chan and Malik [3] obtained a generalization of (5) by considering the lacunary type of polynomials and obtained the following:

Theorem A: *Let* $P(z) := \sum_{j=\mu}^{n} a_j z^j$, $1 \leq \mu \leq n$ *be a polynomial of degree n having all its zeros in* $|z| \geq k, k \geq 1$, *then*

$$\max_{|z|=1}|P'(z)| \leq \frac{n}{1+k^{\mu}}\max_{|z|=1}|P(z)|. \tag{7}$$

The result is best possible and extremal polynomial is $P(z) = \left(z^{\mu} + k^{\mu}\right)^{\frac{n}{\mu}}$, where n is a multiple of μ.

The next result was proved by Pukhta [8], who infact proved:

Theorem B: *Let* $P(z) := \sum_{j=\mu}^{n} a_j z^j$, $1 \leq \mu \leq n$ *be a polynomial of degree n having all its zeros in,* $|z| \geq k, k \geq 1$ *then*

$$\max_{|z|=1}|P'(z)| \leq \frac{n}{1+k^{\mu}}\left\{\max_{|z|=1}|P(z)| - \min_{|z|=k}|P(z)|\right\}. \tag{8}$$

The result is best possible and extremal polynomial is $P(z) = \left(z^{\mu} + k^{\mu}\right)^{\frac{n}{\mu}}$; where n is a multiple of μ.

Bidkham and Dewan [2] obtained a generalization of (5) by proving the following result:

Theorem C: *Let P(z) be a polynomial of degree n having all its zeros in* $|z| \geq k, k \geq 1$, *then for* $1 \leq R \leq k$

$$\max_{|z|=1}|P'(z)| \geq \frac{n(R+k)^{n-1}}{(1+k)^{n}}\max_{|z|=1}|P(z)|. \tag{9}$$

The result is best possible and equality holds for P(z) = (z + k)n.

In this paper, we prove the following generalization as well as an improvement of Theorem C by considering the sth derivative of P(z).

Theorem 1: If $P(z) := \sum_{j=\mu}^{n} a_j z^j$, $n>3$, $1 \leq \mu \leq n$ is a polynomial of degree n having all its zeros in $|z| \geq k, k \geq 1$, then for $1 \leq R \leq k$, and $1 \leq s \leq n$

$$\max_{|z|=1}\left|P^s(z)\right| \leq \frac{n(n-1)\dots(n-s+1)}{R^s+k^s}\left\{\frac{\left(R^{n-1}-1\right)\left(n|a_0|+\mu|a_\mu|k^{\mu+1}\right)}{+n|a_0|\left(1+k^{\mu+1}\right)+\mu|a_\mu|\left(k^{\mu+1}+k^{2\mu}\right)} - \frac{2}{(n+1)}|P'(0)|\left(\frac{\left(R^n-1\right)}{n}-(R-1)\right)\right\}$$

$$-|P''(0)|\left[\left(\frac{\left(R^n-1\right)-n(R-1)}{n(n-1)}\right)-\left(\frac{\left(R^{n-2}-1\right)-(n-2)(R-1)}{(n-2)(n-3)}\right)\right]$$

$$-\frac{n(n-1)\dots(n-s+1)}{R^s+k^s}\min_{|z|=k}|P(z)|.$$

LEMMAS

For the proof of above theorem, we need the following lemmas. The first result is due to Qazi [9, Lemma 1].

Lemma 1: *If* $P(z) := a_0 + \sum_{j=\mu}^{n} a_j z^j$, $1 \leq \mu \leq n$ *is a polynomial of degree n having all its zeros in* $|z| \geq k, k \geq 1$, *then*

$$\max_{|z|=1}\left|P^s(z)\right| \leq n\left\{\frac{n|a_0|+\mu|a_\mu|k^{\mu+1}}{n|a_0|\left(1+k^{\mu+1}\right)+\mu|a_\mu|\left(k^{\mu+1}+k^{2\mu}\right)}\right\}\max_{|z|=1}|P(z)|.$$

The next lemma is due to Dewan, Kour and Mir [4].

Lemma 2: *Let* $P(z) := \sum_{j=0}^{n} a_j z^j$ *be a polynomial of degree n, then for* $R \geq 1$*;*

$$\max_{|z|=1}|P(z)| \leq R^n \max_{|z|=1}|P(z)| - 2\frac{(R^n-1)}{n+2}|P(0)| - \left[\frac{(R^n-1)}{n} - \frac{(R^{n-2}-1)}{n-2}\right]|P'(0)|, n > 2 \tag{10}$$

And

$$\max_{|z|=1}|P(z)| \leq R^n \max_{|z|=1}|P(z)| - \frac{(R-1)}{2}\left[\begin{matrix}(R+1)|P(0)| \\ +(R-1)|P'(0)|\end{matrix}\right], n = 2. \tag{11}$$

Lemma 3: *If* $P(z) := a_0 + \sum_{j=\mu}^{n} a_j z^j,\ n > 3,\ 1 \leq \mu \leq n$ *is a polynomial of degree n having all its zeros in* $|z| \geq k, k \geq 1$*, then for* $R \geq 1, 0 \leq \theta \leq 2\pi$*, we have*

$$|P'(Re^{i\theta})| \leq n\left\{\frac{nR^{n-1}\left(n|a_0| + \mu|a_\mu|k^{\mu+1}\right)}{n|a_0|\left(1+k^{\mu+1}\right) + \mu|a_\mu|\left(k^{\mu+1}+k^{2\mu}\right)}\right\}\max_{|z|=1}|P(z)| - 2\frac{(R^{n-1}-1)}{n+1}|P'(0)| - \left[\frac{(R^{n-1}-1)}{n-1} - \frac{(R^{n-3}-1)}{n-3}\right]|P''(0)|$$

Proof of Lemma 3: Since P(z) is a polynomial of degree n > 3, the polynomial P′(z) is of degree n ≥ 3, hence on applying inequality (10) of Lemma 2 to the polynomial P′(z), we obtain

$$\left|P'\left(\mathrm{Re}^{i\theta}\right)\right| \le R^{n-1}\max_{|z|=1}\left|P(z)\right|$$
$$-2\frac{\left(R^{n-1}-1\right)}{n+1}\left|P'(0)\right| - \left[\frac{\left(R^{n-1}-1\right)}{n-1} - \frac{\left(R^{n-3}-1\right)}{n-3}\right]\left|P''(0)\right|$$

This proves Lemma 3.

Lemma 4: *If* $P(z) := a_0 + \sum_{j=\mu}^{n} a_j z^j, \quad n \; > \; 3, \; 1 \le \mu \le n$ *is a polynomial of degree n having all its zeros in* $|z| \ge k, k \ge 1$, *then for* $R \ge 1$,

$$\max_{|z|=R}\left|P(z)\right| \le$$
$$\left\{\frac{\begin{array}{c}\left(R^{n-1}-1\right)\left(n\left|a_0\right| + \mu\left|a_\mu\right|k^{\mu+1}\right) \\ +n\left|a_0\right|\left(1+k^{\mu+1}\right) + \mu\left|a_\mu\right|\left(k^{\mu+1}+k^{2\mu}\right)\end{array}}{n\left|a_0\right|\left(1+k^{\mu+1}\right) + \mu\left|a_\mu\right|\left(k^{\mu+1}+k^{2\mu}\right)}\right\}\max_{|z|=1}\left|P(z)\right|$$
$$-\frac{2}{(n+1)}\left|P'(0)\right|\left\{\frac{R^n-1}{n} - (R-1)\right\}$$
$$-\left|P''(0)\right|\left[\left(\frac{\left(R^n-1\right)-n(R-1)}{n(n-1)}\right) - \left(\frac{\left(R^{n-2}-1\right)-(n-2)(R-1)}{(n-2)(n-3)}\right)\right]$$

Proof of Lemma 4: For each $0 \le \theta \le 2\pi$ and for $R \ge 1$ we have

$$P\left(\mathrm{Re}^{i\theta}\right) - P\left(\mathrm{e}^{i\theta}\right) = \int_1^R \mathrm{e}^{i\theta}\, P'\left(r\,\mathrm{e}^{i\theta}\right)dr$$

Hence

$$\left|P\left(\mathrm{Re}^{i\theta}\right) - P\left(\mathrm{e}^{i\theta}\right)\right| = \int_1^R P'\left(r\,\mathrm{e}^{i\theta}\right)dr, \tag{12}$$

which when combined with Lemma 3, gives

$$\left|P\left(Re^{i\theta}\right)-P\left(e^{i\theta}\right)\right| = \int_1^R \left\{ \begin{array}{l} \dfrac{nr^{n-1}\left(n|a_0|+\mu|a_\mu|k^{\mu+1}\right)}{n|a_0|\left(1+k^{\mu+1}\right)+\mu|a_\mu|\left(k^{\mu+1}+k^{2\mu}\right)}\max_{|z|=1}|P(z)| \\ -2\dfrac{\left(r^{n-1}-1\right)}{n+1}|P'(0)| \\ -\left[\dfrac{\left(r^{n-1}-1\right)}{n-1}-\dfrac{\left(r^{n-3}-1\right)}{n-3}\right]|P''(0)| \end{array} \right\} dr,$$

which gives

$$\max_{|z|=R}|P(z)| \le \frac{\left(R^{n-1}-1\right)\left(n|a_0|+\mu|a_\mu|k^{\mu+1}\right) + n|a_0|\left(1+k^{\mu+1}\right)+\mu|a_\mu|\left(k^{\mu+1}+k^{2\mu}\right)}{n|a_0|\left(1+k^{\mu+1}\right)+\mu|a_\mu|\left(k^{\mu+1}+k^{2\mu}\right)}\max_{|z|=1}|P(z)|$$

$$-\frac{2}{(n+1)}|P'(0)|\left\{\frac{R^n-1}{n}-(R-1)\right\}$$

$$-|P''(0)|\left[\left(\frac{\left(R^n-1\right)-n(R-1)}{n(n-1)}\right)-\left(\frac{\left(R^{n-2}-1\right)-(n-2)(R-1)}{(n-2)(n-3)}\right)\right]$$

Hence the proof.

Lemma 5: *If P(z) is a polynomial of degree n having all its zeros in* $|z| \ge k, k \ge 1$, *then for* $1 \le s \le n$.

$$\max_{|z|=1}\left|P^{s}(z)\right| \le \frac{n(n-1)...n(n-s+1)}{1+k^{s}}$$
$$\left\{\max_{|z|=1}|P(z)| - \max_{|z|=k}|P(z)|\right\}.$$

This Lemma is due to Govil [5].

Proof of Theorem 1: Since P(z) has all its zeros in $|z| \ge k, k \ge 1$ and if $1 \le R \le k$, then G(z) = P(Rz) has all its zeros in $|z| \ge \frac{k}{R}, \frac{k}{R} \ge 1$, therefore by applying Lemma 5 to G(z), we obtain

$$\max_{|z|=1}\left|G^{s}(z)\right| \le \frac{n(n-1)...n(n-s+1)}{1+\left(\frac{k}{R}\right)^{s}}$$
$$\left\{\max_{|z|=1}|G(z)| - \max_{|z|=\frac{k}{R}}|G(z)|\right\}.$$

which implies

$$R^{s}\max_{|z|=1}\left|G^{s}(Rz)\right| \le \frac{n(n-1)...n(n-s+1)}{1+\left(\frac{k}{R}\right)^{s}}$$
$$\left\{\max_{|z|=1}|G(Rz)| - \max_{|z|=\frac{k}{R}}|G(Rz)|\right\}.$$

which is equivalent to

$$\max_{|z|=R}\left|G^{s}(z)\right| \le \frac{n(n-1)...n(n-s+1)}{R+k^{s}}$$
$$\left\{\max_{|z|=R}|G(z)| - \max_{|z|=k}|G(z)|\right\}. \tag{13}$$

Inequality (13) in conjunction with Lemma 4 yields

$$\max_{|z|=R}\left|P^{s}(z)\right| \le \frac{n(n-1)\dots(n-s+1)}{R^{s}+k^{s}}$$

$$\left\{\frac{\left(R^{n-1}-1\right)\left(n|a_0|+\mu|a_\mu|k^{\mu+1}\right) + n|a_0|\left(1+k^{\mu+1}\right)+\mu|a_\mu|\left(k^{\mu+1}+k^{2\mu}\right)}{n|a_0|\left(1+k^{\mu+1}\right)+\mu|a_\mu|\left(k^{\mu+1}+k^{2\mu}\right)}\right\}\max_{|z|=1}|P(z)|$$

$$-\frac{2}{(n+1)}|P'(0)|\left\{\frac{R^{n}-1}{n}-(R-1)\right\}$$

$$-|P''(0)|\left[\left(\frac{\left(R^{n}-1\right)-n(R-1)}{n(n-1)}\right)-\left(\frac{\left(R^{n-2}-1\right)-(n-2)(R-1)}{(n-2)(n-3)}\right)\right]$$

$$-\frac{n(n-1)\dots(n-s+1)}{R^{s}+k^{s}}\min_{|z|=k}|P(z)|.$$

The proof of Theorem 1 is completed.

REFERENCES

1. A. Aziz and Q.M. Dawood, Inequalities for a polynomial and its derivative, J. Approx. Theory, 54 (1988), No. 3, 306-313.
2. M. Bidkham and K. K. Dewan, Inequalities for a polynomial and its derivative, J. Math. Anal. Appl., Vol. 166 (1992), 319-324.
3. T. N. Chan and M. A. Malik, On Erd os-Lax Theorem, Proc. Indian Acad. Sci. (Math. Sci.), 92 (3) (1983), 191-193.
4. K. K. Dewan, Jagjeet Kaur and Abdullah Mir, Inequalities for the derivative of a polynomial, J. Math. Anal. Appl., 269 (2002), 489-499.
5. N. K. Govil, Some inequalities for derivatives of polynomials, J. Approx. Theory, 66 (1) (1991), 29-35.
6. P. D. Lax, Proof of a conjecture of P. Erd os on the derivative of a polynomial, Amer. Math. Soc., Bulletin, 50 (1944), 509-513.

7. M. A. Malik, On the derivative of a polynomial, J. London Math. Soc., 2 (1) (1969), 57-60.

8. M. S. Pukhta Extremal Problems for Polynomials and on Location of Zeros of Polynomials, Ph. D Thesis, Jamia Millia Islamia, New Delhi (1995).

9. M. A. Qazi, On the maximum modulus of polynomials, Proc. Amer. Math. Soc., 115(1992), 337-343.

10. P. Turán, Uber die Ableitung von Polynomen, Compositio Mathematica, 7 (1939), 89-95 (German).

12

HE-LAPLACE METHOD FOR THE SOLUTION OF TWO-POINT BOUNDARY VALUE PROBLEMS

Hradyesh Kumar Mishra*

[1] Department of Mathematics, Jaypee University of Engineering & Technology, Guna-473226(M.P), India

ABSTRACT

The boundary value problems of ordinary differential equations play an important role in many fields. Here, we implement the He-Laplace method for the solution of linear and nonlinear two-point boundary value problems. The aim of this paper is to compare the performance of the He-Laplace method with shooting method. As a result, for the same number of terms, our method provides relatively more accurate results with rapid convergence than other methods.

INTRODUCTION

There are many new analytical approximate methods to solve two-point boundary value and initial value problems in the literature. These problems generally arise frequently in many areas of science and engineering, for example, fluid mechanics, quantum mechanics, optimal control, chemical-reactor theory, aero-dynamics, reaction-diffusion process, geophysics etc. Different numerical methods have been proposed by various researchers. Among these are the Adomian decomposition method (ADM) [1, 2, 3, 23, 35, 36, 37], Homotopy perturbation method HPM [7-17,19,20,32,33,39], He-Laplace method [21Journal of Applied Mathematics Vol. 2012,2012, pp.1-16." , 22, 23, 26, 27, 28, 29, 30]. There are many techniques available for the numerical solution of two-point boundary value problems for ordinary differential equations [18, 25] and references therein. However there is no unified method to handle all types of nonlinear problems. Mohsen and El-Gamel [31] studied the performance of the collocation and Galerkin methods using Sinc basis functions for solving linear and nonlinear second-order two-point boundary value problems. The Sinc-collocation method is used to solve a system of second-order boundary value problems [5].

In this paper, we implement the He-Laplace method for solving the two-point boundary value problem of the form:

$$y'' = f(x, y, y'), \quad a < x < b \tag{1}$$

with boundary conditions

$$y(a) = \alpha \ , \quad y(b) = \beta \tag{2}$$

where f is continuous on the set $D = \{\ (x, y, y') | a \leq x \leq b,\ y, y' \in R\}$, a, b, α, β are constants. The nonlinear shooting method [22] and He-Laplace method are used to find the solutions of the nonlinear two-point boundary value problems (1)-(2). In this paper, we present some numerical examples including linear as well as nonlinear boundary value problems. The main purpose of this paper is to compare the solutions obtained by He-Laplace method with Extended Adomian decomposition method (EADM) and Shooting method.

This paper contains basic idea of homotopy perturbation method and He-Laplace method in section 2, Numerical examples in section 3 and conclusion in section 4 respectively.

BASIC IDEA OF HOMOTOPY PERTURBATION METHOD

Consider the following nonlinear differential equation

$$A(y) - f(r) = 0 \ , \quad r \in \Omega \tag{3}$$

with the boundary conditions of

$$B\left(y, \frac{\partial y}{\partial n}\right) = 0, \quad r \in \Gamma, \tag{4}$$

where A, B, $f(r)$ and Γ are a general differential operator, a boundary operator, a known analytic function and the boundary of the domain Ω, respectively.

The operator A can generally be divided into a linear part L and a nonlinear part N. Eq. (3) may therefore be written as:

$$L(y)+N(y)-f(r)=0, \tag{5}$$

By the homotopy technique, we construct a homotopy $v(r,p):\Omega\times[0,1]\to R$ which satisfies:

$$H(v,p)=(1-p)\left[L(v)-L(y_0)\right]+p\left[A(v)-f(r)\right]=0$$

or

$$H(v,p)=L(v)-L(y_0)+pL(y_0)+p\left[N(v)-f(r)\right]=0$$

where $p\in[0,1]$ is an embedding parameter, while y_0 is an initial approximation of Eq.(1), which satisfies the boundary conditions. Obviously, from above equations,

we will have:

$$H(v,0)=L(v)-L(y_0)=0, \tag{6}$$

$$H(v,1)=A(v)-f(r)=0, \tag{7}$$

The changing process of p from zero to unity is just that of $v(r,p)$ from y_0 to $y(r)$. In topology, this is called deformation, while $L(v)-L(y_0)$ and $A(v)-f(r)$ are called homotopy. If the embedding parameter p is considered as a small parameter, applying the classical perturbation technique, we can assume that the solution of Eqs.(6) and (7) can be written as a power series in p:

$$v=v_0+pv_1+p^2v_2+p^3v_3+\ldots\ldots\infty \tag{8}$$

Setting $p=1$ in Eq.(8), we have

$$y=\lim_{p\to 1}v=v_0+v_1+v_2+\ldots\ldots \quad (9) \tag{9}$$

The combination of the perturbation method and the homotopy method is called the HPM, which eliminates the drawbacks of the traditional perturbation methods while keeping all its advantages. The series (9) is convergent for most cases. However, the convergent rate depends on the nonlinear operator $A(v)$. Moreover, He [35] made the following suggestions:

(1) The second derivative of $N(v)$ with respect to v must be small because the parameter may be relatively large, i.e. $p \to 1$.
(2) The norm of $L^{-1}\left(\frac{\partial N}{\partial v}\right)$ must be smaller than one so that the series converges.

HE-LAPLACE METHOD

Consider the following nonlinear differential equation (IVP):

$$y'' + p_1 y' + p_2 y + p_3 f(y) = f(x) \tag{10}$$

$$y(0) = \alpha, \quad y'(0) = \beta \tag{11}$$

where $p_1, p_2, p_3, \alpha, \beta$ are constants. f(y) is a nonlinear function and f(x) is the source term. Taking Laplace transformation (denoted throughout this paper by L) on both side of Eq (10), we have

$$L[y''] + L[p_1 y'] + L[p_2 y] + L[p_3 f(y)] = L[f(x)] \tag{12}$$

By using linearity of Laplace transformation, the result is

$$L[y''] + p_1 L[y'] + p_2 L[y] + p_3 L[f(y)] = L[f(x)] \tag{13}$$

Applying the formula on Laplace transform, we obtain

$$\begin{aligned} &s^2 L[y] - sy(0) - y'(0) + p_1 \{sL[y] - y(0)\} \\ &+ p_2 L[y] + p_3 L[f(y)] = L[f(x)] \end{aligned} \tag{14}$$

Using initial conditions in Eq.(14), we have

$$\begin{aligned} &(s^2 + p_1 s) L[y] \\ &= \alpha s + \beta + \alpha p_1 - p_2 L[y] - p_3 L[f(y)] + L[f(x)] \end{aligned} \tag{15}$$

Or

$$L[y] = \frac{(\alpha s + \beta + \alpha p_1)}{(s^2 + p_1 s)} - \frac{p_2}{(s^2 + p_1 s)} L[y] - \frac{p_3}{(s^2 + p_1 s)} L[f(y)] + L[f(x)] \tag{16}$$

Taking inverse Laplace transform, we have

$$y(x) = F(x) - L^{-1}\left(\frac{p_2}{s^2 + p_1 s} L[y] \right) - L^{-1}\left(\frac{p_3}{s^2 + p_1 s} L[f(y)] \right) \tag{17}$$

where $F(x)$ represents the term arising from the source term and the prescribed initial conditions.

Now, we apply homotopy perturbation method [35],

$$y(x) = \sum_{n=0}^{\infty} p^n y_n(x) \tag{18}$$

where the term y_n are to recursively calculated and the nonlinear term $f(y)$ can be decomposed as

$$f(y) = \sum_{n=0}^{\infty} p^n H_n(y) \tag{19}$$

for some He's polynomial H_n (see [[21, 22] and their referencein]) that are given by

$$H_n(y_0, y_1, y_2, \ldots\ldots\ldots\ldots, y_n) = \frac{1}{n!} \frac{\partial^n}{\partial p^n} \left[f\left(\sum_{i=0}^{\infty} p^i y_i \right) \right]_{p=0} \qquad n = 0, 1, 2, 3 \ldots\ldots\ldots\ldots$$

Substituting Eqs.(18) and (19) in (17), we get

$$\sum_{n=0}^{\infty} p^n y_n(x) = F(x) - p\begin{pmatrix} L^{-1}\left\{\frac{p_2}{(s^2+p_1 s)} L\left[\sum_{n=0}^{\infty} p^n y_n(x)\right]\right\} \\ +L^{-1}\left\{\frac{p_3}{(s^2+p_1 s)} L\left[\sum_{n=0}^{\infty} p^n H_n(y)\right]\right\} \end{pmatrix} \quad (20)$$

which is the coupling of the Laplace transformation and the homotopy perturbation method using He's polynomials. Comparing the coefficient of like powers of p, the following approximations are obtained:

$$p^0: \quad y_0(x) = F(x),$$

$$p^1: \quad y_1(x) = -\begin{pmatrix} L^{-1}\left\{\frac{p_2}{(s^2+p_1 s)} L[y_0(x)]\right\} \\ +L^{-1}\left\{\frac{p_3}{(s^2+p_1 s)} L[H_0(y)]\right\} \end{pmatrix}$$

$$p^2: \quad y_2(x) = -\begin{pmatrix} L^{-1}\left\{\frac{p_2}{(s^2+p_1 s)} L[y_1(x)]\right\} \\ +L^{-1}\left\{\frac{p_3}{(s^2+p_1 s)} L[H_1(y)]\right\} \end{pmatrix} \quad (21)$$

$$p^3: \quad y_3(x) = -\begin{pmatrix} L^{-1}\left\{\frac{p_2}{(s^2+p_1 s)} L[y_2(x)]\right\} \\ +L^{-1}\left\{\frac{p_3}{(s^2+p_1 s)} L[H_2(y)]\right\} \end{pmatrix}.$$

NUMERICAL EXAMPLES

In this section, two examples are used to illustrate performance of the He-Laplace method. For the sake of comparison with other methods, we consider the same examples as in [6, 24, 34]. In the process of Computation, all the symbolic and numerical computations are performed by using Mathematica.

Example 4.1. Consider the linear two-point boundary value problem of the form [24, 34]

$$y'' = y + \cos(x), \quad 0 < x < 1, \tag{22}$$

Subject to the boundary conditions

$$y(0) = 1 \;, \quad y(1) = 1 \tag{23}$$

The exact solution of the problem is

$$y(x) = \frac{-3\cosh(1) + 3\sinh(1) + \cos(1) + 2}{4\sinh(1)}\exp(x) + \frac{3\cosh(1) + 3\sinh(1) - \cos(1) - 2}{4\sinh(1)}\exp(-x) - \frac{\cos(x)}{2} \tag{24}$$

By applying the aforesaid method subject to the initial condition, we have

$$y(s) = \frac{s+\beta}{s^2} + \frac{1}{s(s^2+1)} + \frac{1}{s^2}L[y] \tag{25}$$

where $y'(0) = \beta$.

The inverse of the Laplace transform implies that

$$y(x) = 1 + \beta x + 1 - \cos x + L^{-1}\left[\frac{1}{s^2}L[y]\right] \tag{26}$$

Now, we apply the homotopy perturbation method, we have

$$\sum_{n=0}^{\infty} p^n y_n(x) = 2 + \beta x - \cos x + p\left(L^{-1}\left[\frac{1}{s^2}L[y]\right]\right) \tag{27}$$

Comparing the coefficient of like powers of p, we have

$$\begin{aligned}
p^0 &: \quad y_0(x) = 2 + \beta x - \cos x, \\
p^1 &: \quad y_1(x) = L^{-1}\left[\frac{1}{s^2} L[y_0(x)]\right] \\
&= L^{-1}\left[\frac{1}{s^3} + \frac{\beta}{s^4} + \frac{1}{s^3(s^2+1)}\right] \\
&= x^2 + \frac{\beta x^3}{6} - 1 + \cos x, \\
p^2 &: \quad y_2(x) = L^{-1}\left[\frac{1}{s^2} L[y_1(x)]\right] \\
&= -\cos x + \frac{\beta x^5}{120} + \frac{x^4}{12} - \frac{x^2}{2} + 1,
\end{aligned} \tag{28}$$

Proceeding in a similar manner, we have

$$p^3: \quad y_3(x) = \frac{\beta x^7}{5040} + \frac{x^2}{2} - \frac{x^4}{24} + \frac{x^6}{360} + \cos x - 1,$$

$$p^4: \quad y_4(x) = \frac{\beta x^9}{362880} - \cos x - \frac{x^2}{2} + \frac{x^4}{24} - \frac{x^6}{720} + \frac{x^8}{20160} + 1,$$

$$p^5: \quad y_5(x) = \frac{\beta x^{11}}{39916800} + \cos x + \frac{x^2}{2} - \frac{x^4}{24} + \frac{x^6}{720} - \frac{x^8}{40320} + \frac{x^{10}}{1814400} - 1,$$

So that the solution $y(x)$ is given by

$$\begin{aligned}
y(x) &= y_0 + y_1 + y_2 + y_3 + y_4 + y_5 \\
&= \beta x + 1 + \frac{\beta x^3}{6} + \frac{\beta x^5}{120} + \frac{x^4}{24} + \frac{\beta x^7}{5040} + \frac{x^6}{360} \\
&+ \frac{\beta x^9}{362880} + \frac{x^8}{40320} + \frac{\beta x^{11}}{39916800} + \frac{x^{10}}{1814400}.
\end{aligned} \tag{29}$$

TABLE 1. Absolute error for example 4.1

x	$\left|y(x)_{exact} - y_5(x)\right|$ in EADM	$\left|y(x)_{exact} - y_5(x)\right|$ in He-Laplace method
1/8	4.37×10^{-7}	0.2×10^{-9}
2/8	8.07×10^{-7}	0.4×10^{-9}
3/8	1.05×10^{-6}	0.5×10^{-9}
4/8	1.14×10^{-6}	0.6×10^{-9}
5/8	1.06×10^{-6}	0.13×10^{-8}
6/8	8.07×10^{-7}	0.12×10^{-8}
7/8	4.37×10^{-7}	0.11×10^{-8}

The approximation must satisfy the boundary conditions. Imposing the boundary condition of Eq.(23) at x = 1 on Eq.(29), we can obtain the value of the parameter $\beta = -0.8887582849$.

Table 1 shows the comparison between exact solution and the series solution obtained by the exponential fitting method (EFM) in [34] and the HLHPM. We compare only five iterations of both the methods. It is found that for the same number of terms, the HLHPM yields relatively more accurate results than other methods. It is obvious that evaluation of more components of u(x) will reasonably improve the accuracy of series solution by using the HLHPM.

Example 4.2. Consider the nonlinear two-point boundary value problem of the form [17, 22]

$$u'' = u^2 + 2\pi^2 \cos(2\pi x) - \sin^4(\pi x), \quad 0 < x < 1, \tag{30}$$

with the boundary conditions

$$u(0) = 0, \quad u(1) = 0. \tag{31}$$

The exact solution of the problem is

$$u(x) = \sin^2(\pi x).$$

By applying the aforesaid method subject to the initial condition, we have

$$y(s) = \frac{s+\beta}{s^2} + \frac{2\pi^2 s}{s^2(s^2+4\pi^2)} - \frac{24\pi^4}{s^3(s^2+4\pi^2)(s^2+16\pi^2)} + \frac{1}{s^2}L\left[y^2\right] \tag{32}$$

where $y'(0) = \beta$.

Taking inverse Laplace transform, we have

$$y(x) = L^{-1}\left[\frac{s+\beta}{s^2} + \frac{2\pi^2}{s(s^2+4\pi^2)} - \frac{24\pi^4}{s^3(s^2+4\pi^2)(s^2+16\pi^2)} + \frac{1}{s^2}L\left[y^2\right]\right]$$

$$y(x) = 1 + \beta x + \frac{1}{2}(1-\cos(2\pi x)) + \left[\frac{15-24\pi^2x^2-16\cos(2\pi x)+\cos(4\pi x)}{128\pi^2}\right] + L^{-1}\left[\frac{1}{s^2}L\left[y^2\right]\right] \tag{33}$$

Now, we apply homotopy perturbation method [35],

$$y(x) = \sum_{n=0}^{\infty} p^n y_n(x) \tag{34}$$

where the term y_n are to recursively calculated and the nonlinear term f(y) can be decomposed as

$$f(y) = \sum_{n=0}^{\infty} p^n H_n(y) \tag{35}$$

For some He's polynomial H_n that are given by

$$H_n(y_0, y_1, y_2, ...y_n) = \frac{1}{n!}\frac{\partial^n}{\partial p^n}\left[f\left(\sum_{i=0}^{\infty} p^i y_i \right) \right]_{p=0} \tag{36}$$

$$n = 0, 1, 2, 3, ...$$

$$\sum_{n=0}^{\infty} p^n y_n(x) = f(x) + p\left(L^{-1}\left\{ \frac{1}{s^2} L\left[\sum_{n=0}^{\infty} p^n H_n(y) \right] \right\} \right) \tag{37}$$

p^0: $\quad y_0(x) = f(x)$

p^1: $\quad y_1(x) = L^{-1}\left\{ \frac{1}{s^2} L[H_0(y)] \right\}$

p^2: $\quad y_2(x) = L^{-1}\left\{ \frac{1}{s^2} L[H_1(y)] \right\}$

p^3: $\quad y_3(x) = L^{-1}\left\{ \frac{1}{s^2} L[H_2(y)] \right\}$

Preceding the steps as in the previous example, we can obtain the value of β and consequently determine the approximate solution with seven iterations $y_7(x)$. The absolute errors between the exact solution and the results obtained by the shooting method [6], EADM and the HLHPM are compared in Table 2. It is worth

TABLE 2. Absolute error for example 4.2

x	$\lvert y(x)_{exact} - y_7(x) \rvert$ in EADM	$\lvert y(x)_{exact} - y_7(x) \rvert$ in He-Laplace method
0.1	$6.9x10^{-7}$	$-0.10x10^{-9}$
0.2	$1.3x10^{-6}$	$0.5x10^{-9}$
0.3	$1.9x10^{-6}$	$0.5x10^{-9}$
0.4	$2.3x10^{-6}$	$-0.1x10^{-9}$
0.5	$2.5x10^{-6}$	$0.1x10^{-9}$
0.6	$2.3x10^{-6}$	$0.6x10^{-9}$
0.7	$1.9x10^{-6}$	$0.6x10^{-9}$
0.8	$1.3x10^{-6}$	$0.7x10^{-9}$
0.9	$6.9x10^{-7}$	$0.95x10^{-9}$

to mention that shooting method requires several numerical procedures in obtaining approximations. It is noted that $y_7(x)$ give much better numerical approximations than the results by the EADM. It is interesting to point out that He's polynomial plays an important role in handling nonlinear terms and making the solution procedure in a systematic way.

CONCLUSION

In this paper, we have applied He-Laplace method for obtaining approximate solutions of linear and nonlinear second-order two-point boundary value problems. Further, we have compared the performance of the He-Laplace method with extended adomian decomposition method. Numerical examples are presented. On the basis of these examples, for the same number of terms, He-Laplace method yields much better results than the other methods.

REFERENCES

1. G.Adomian, A revieal of the decomposition method and some recent results for nonlinear equation, Mathematical and Computer Modeling 13(7) (1992) 17-43.
2. G.Adomian, A revieal of the decomposition method in applied mathematics, Journal of Mathematical Analysis and application 135(1988) 501-544.
3. G.Adomian, Solution frontier Problems of Physics: The Decomposition method, Kluwer Dordrecht, 1994.
4. C.Chun, R.Sakthivel, Homotopy perturbation technique for solving two point boundary value problems- comparison with other methods, Computer physics communications, 181(2010) 1021-1024.
5. M.Dehghan, A Saadatmandi, the numerical solution of a nonlinear system of second order boundary value problems using the Sinc-Collocation method, Mathematical and Computer Modeling 46(2007) 1434-1441.
6. S.N. Ha, A nonlinear shooting method for two-point boundary value problems, Computers and Mathematics with Applications 42(2001) 1411-1420.
7. J.H.He, Homotopy perturbation method: A new nonlinear analytical technique, Applied Mathematics and Computation, Vol. 135, 2003, pp.73-79.
8. J.H.He, Homotopy, Comparion of homotopy perturbation method and homotopy analysis method, Applied Mathematics and Computation, Vol. 156, 2004, pp. 527-539.
9. J.H.He, Homotopy, The homotopy perturbation method for nonlinear oscillators with discontinueties, Applied Mathematics and Computation, Vol. 151, 2004, pp. 287-292.

10. J.H.He, Recent developments of the homotopy perturbation method, Topological methods in Non-linear Analysis, Vol. 31, 2008, pp. 205-209.

11. J.H.He, New Interpretation of homotopy perturbation method, International Journal of Modern Physics, Vol. 20, 2006, pp. 2561-2568.

12. J.H.He, A coupling method of homotopy technique and a perturbation technique for nonlinear problems, International Journal of Nonlinear Mechanics, Vol. 35, 2000, pp. 37-43.

13. J.H.He, Variational iteration method for autonomous ordinary differential systems, Applied Mathematics and Computation, Vol. 114, 2000, pp. 115-123.

14. J.H.He, Homotopy perturbation technique, Computer methods in Applied Mechanics and Engineering, Vol. 178, 1999, pp. 257-262.

15. J.H.He, A simple perturbation approach to Blasius equation, Applied Mathematics and Computation, Vol.140, 2003, pp. 217-222.

16. J.H.He, Application of homotopy perturbation method to nonlinear wave equation, Chaos, Solitons, Fractals, Vol. 26, 2005, pp. 295-300.

17. J.H.He, Homotopy perturbation method for solving boundary value problem, Physics Letter A, Vol. 350, 2006, pp. 87-88.

18. M.T.Heath, Scientific Computing; An Introductory Survey, McGraw-Hill, New York 2002.

19. E. Hesameddini, H. Latifizadeh, An optimal choice of initial solutions in the homotopy perturbation method, International Journal of Nonlinear Sciences and Numerical Simulation, Vol. 10, 2009, pp.1389-1398.

20. E. Hesameddini, H. Latifizadeh, A new vision of the He's homotopy perturbation method, International Journal of Nonlinear Sciences and Numerical Simulation, Vol.10, 2009, pp.1415-1424.

21. H.K. Mishra, A.K. Nagar, He-Laplace method for linear and nonlinear partial differential equations, Journal of Applied Mathematics Vol. 2012,2012, pp.1-16.

22. H.K. Mishra, He-Laplace Method for Special Nonlinear Partial Differential Equations, Mathematical Theory and Modeling Vol.3 (6), 2013, pp. 113-117.

23. S. Islam, Y. Khan, N. Faraz, F. Austin, Numerical solution of logistic differential equations by using the Laplace decomposition method, World Applied Sciences Journal, Vol. 8, 2010, pp. 1100-1105.

24. B.Jang, Two point boundary value problems by the extended Adomian decomposition method, Journal of Computational and Applied Mathematics 219(2008)253-262.

25. K.Keller, Numerical solution of two-point boundary value problems, SIAM, Philadelphia, 1976.

26. Y.Khan, F.Austin, Application of the Laplace decomposition method to nonlinear homogeneous and non-homogeneous advection equations, Zeitschrift fuer Naturforschung, Vol. 65 a, 2010, pp.1-5.

27. Yasir Khan, Qingbiao Wu, Homotopy perturbation transform method for nonlinear equations using He's polynomials, Computers and Mathematics with Applications, Vol.61, 2011, pp.1963-1967.

28. Yasir khan, An effective modification of the Laplace decomposition method for nonlinear equations, International Journal of Nonlinear Sciences and Numerical Simulation, Vol. 10, 2009, pp. 1373-1376.

29. S.A.Khuri, A Laplace decomposition algorithm applied to a class of nonlinear differential equations, Journal of Applied Mathematics,Vol.1, 2001, pp.141-155.

30. M.Madani, M.Fathizadeh, Homotopy perturbation algorithm using Laplace transformation, Nonlinear Science Letters A, Vol. 1, 2010, pp.263-267.

31. A.Mohsen, M.El-Gamel, On the Galerkin and Collocation methods for two-point boundary value problems using Sinc bases, Computers and Mathematics with applications 56(2008)930-941.

32. M.Rafei,D.D.Ganji, Explicit solutions of helmhotz equation and fifth-order KdV equation using homotopy perturbation method, International Journal of Nonlinear Sciences and Numerical Simulation, vol. 7, 2006, pp. 321-328.

33. A.M.Siddiqui, R.Mohmood,Q.K.Ghori,Thin film flow of a third grade fluid on a moving belt by He's homotopy perturbation method, International Journal of Nonlinear Sciences and Numerical Simulation, vol. 7, 2006, pp. 7-14.

34. O.A. Taiwo, Exponential fitting for the solution of two-point boundary value problems with cubic splinecollocation tau method International Journal of Computer Mathematics, 79(2002)299-306.

35. A.M. Waz Waz, A reliable modification of Adomian's decomposition method, Applied Mathematics and computation 92(1998)1-3.

36. A.M.Waz Waz, A composition between Adomian's decomposition method and Taylor series method in the series solution, Applied Mathematics and computation 79(1998)37-44.

37. A.M.Waz Waz, Adomian decomposition method for a reliable treatment of the Bratu type equations, Applied Mathematics and computation 166(2005)652-663.

38. A.M.Waz Waz, Partial Differential Equations; Methods and Applications, Balkema Publishers, The Netherland, 2002.

39. L.Xu, He's homotopy perturbation method for a boundary layer equation in unbounded domain, Computers and Mathematics with Applications, vol. 54, 2007, pp. 1067-1070.

13

SOME IDENTITIES INVOLVING COMMON FACTORS OF K-FIBONACCI AND K-LUCAS NUMBERS

Deepika Jhala[2,*] , G.P.S. Rathore[1], Bijendra Singh[2]

[1] School of Studies in Mathematics, Vikram University, Ujjain (India)

[2] College of Horticulture, Mandsaur (India)

ABSTRACT

Fibonacci sequence stands as a kind of super sequences with fabulous properties. In this paper we present identities involving Common factors of k-Fibonacci and k-Lucas number. Also Binet's formula will employ to this identity.

INTRODUCTION

The Fibonacci sequence and the Lucas sequence are the two shining stars in the vast array of integer sequences. They have fascinated both amateurs and professional mathematicians for centuries, and they continue to charm us with their beauty, abundant applications, and ubiquitous habit of occurring in totally surprising and unrelated places.

The Fibonacci numbers $\{F_n\}_{n\in N}$ are terms of the sequence {0, 1, 1, 2, 3, 5, 8, 13, 21,. ..}, where each term is the sum of the two previous terms, beginning with the values $F_0 = 0 \; and \; F_1 = 1$. The Lucas numbers $\{L_n\}_{n\in N}$ are terms of the sequence {2, 1, 3, 4, 7, 11, 18, 29, 47,...}, where each term is the sum of the two previous terms, beginning with the values $L_0 = 2 \; and \; L_1 = 1$. Many kinds of generalizations of Fibonacci numbers and Lucas numbers have been presented in the literature [6,7,9] One example of this recent generalization is k-Fibonacci numbers [2] and k-Lucas numbers [4], which are defined by the recurrence relation, depend on one real parameter k. Falcon and Plaza [3] defined k- Fibonacci hyperbolic functions and deduced some properties of k- Fibonacci hyperbolic functions related with the analogous identities for the k- Fibonacci numbers. Falcon [4] studied the k- Lucas numbers and proved various properties related with the k- Fibonacci numbers. Thongmoon [11] and Hoggatt [5] defined various identities for the Fibonacci and Lucas numbers. Gupta and Panwar [10], present generalized identities involving common factors of generalized Fibonacci, Jacobsthal and jacobsthal-Lucas numbers and Binet's formula will employ to obtain the identities. There are a lot of identities of Fibonacci and Lucas numbers described in.[1,8] In this paper, we present generalized identities involving common factors of k- Fibonacci, and k- Lucas numbers.

THE K-FIBONACCI AND THE K-LUCAS NUMBERS

Before presenting our main results, we will need to introduce some known results and notations.

The k-Fibonacci numbers have been defined in [2] for any real k as follows:

For any positive real number k, the k-Fibonacci sequence say $\{F_{k,n}\}_{n\in N}$ is define recurrently by

$$F_{k,n+1} = kF_{k,n} + F_{k,n-1} \quad for \ n \geq 1; \tag{1}$$

With initial condition

$$F_{k,0} = 0, \ F_{k,1} = 1 \tag{2}$$

Particular cases of the k-Fibonacci sequence are

If k = 1, the classical Fibonacci sequence is obtained

$$F_0 = 0, F_1 = 1 and \ F_{n+1} = F_n + F_{n-1} \ for \ \ n \geq 1;$$
$$\{F_{k,n}\}_{n\in N} = \{0, 1, 1, 2, 3, 5, 8, ...\}$$

If k = 2, the classical Pell sequence is obtained

$$P_0 = 0, P_1 = 1 and P_{n+1} = 2P_n + P_{n-1} for \ \ n \geq 1;$$
$$\{P_{k,n}\}_{n\in N} = \{0,1,2,12,29,70, ...\}$$

If k = 3, the following sequence is obtained

$$H_0 = 0, H_1 = 1 \text{ and } H_{n+1} = 2H_n + H_{n-1} \text{ for } n \geq 1;$$
$$\{H_{k,n}\}_{n \in N} = \{0, 1, 3, 10, 33, 109, \ldots\ldots\ldots\ldots\}$$

Binet formula for the nth k-Fibonacci numbers is given by

$$F_{k,n} = \frac{r_1^n - r_2^n}{r_1 - r_2} \tag{3}$$

where r_1 , r_2 are the roots of the characteristic equation $r^2 = kr + 1$ and; $r_1 > r_2$; $r_1 = \frac{k + \sqrt{k^2 + 4}}{2}$, $r_2 = \frac{k - \sqrt{k^2 + 4}}{2}$ which gives

$$r_1 + r_2 = k, r_1 r_2 = -1, r_1 - r_2 = \sqrt{k^2 + 4} \tag{4}$$

Similarly for any positive real number k, the k-Lucas sequence [4] say $\{L_{k,n}\}_{n \in N}$ is define recurrently by

$$L_{k,n+1} = kL_{k,n} + L_{k,n-1} \qquad \text{for } n \geq 1 \tag{5}$$

with initial condition

$$L_{k,0} = 2, \; L_{k,1} = k \,. \tag{6}$$

Particular cases of the k-Lucas sequence are

If k = 1, the classical Lucas sequence is obtained

$$L_0 = 2, L_1 = 1 \text{ and } L_{n+1} = L_n + L_{n-1} \text{ for } n \geq 1;$$
$$\{L_{k,n}\}_{n \in N} = \{2, 1, 3, 4, 7, 11, 18, \ldots\ldots\ldots\ldots\}.$$

If k = 2, the classical Pell- Lucas sequence is obtained

$$p_0 = 2, \quad p_1 = 2 \text{ and } p_{n+1} = 2p_n + p_{n-1} \text{ for } n \geq 1;$$
$$\{p_{k,n}\}_{n \in N} \{2, 2, 6, 14, 34, 82, \ldots\ldots\ldots\ldots\}$$

If k = 3, the following sequence is obtained

$$H_0 = 2, H_1 = 3 \text{ and } H_{n+1} = 2H_n + H_{n-1} \text{ for } n \geq 1;$$
$$\{H_{k,n}\}_{n \in N} = \{2, 3, 11, 36, 119, 393, \ldots\ldots\ldots\}.$$

Binet formula for the nth k-Lucas numbers is given by

$$L_{k.n} = r_1^n + r_2^n \tag{7}$$

Where r_1 , r_2 are the roots of the characteristic equation which are given in equation (4).

There are a lot of identities for the common factor of Fibonacci and the Lucas numbers define in [12].

IDENTITIES FOR THE COMMON FACTORS OF K-FIBONACCI AND K-LUCAS NUMBER

In this section we present generalized identities involving common factors of k-Fibonacci and k-Lucas numbers. We shall use the Binet's formula for the k-Fibonacci numbers and k-Lucas numbers for derivation.

Proposition

$$F_{k.4n+p} - F_{k,p} = F_{k,2n} L_{k,2n+p}, \text{ where } n \geq 1 \text{ and } p \geq 0.$$

Proof

$$F_{k,2n} L_{k,2n+p} = \left(\frac{r_1^{2n} - r_2^{2n}}{r_1 - r_2} \right) \left(r_1^{2n+p} + r_2^{2n+p} \right)$$
$$= \left(\frac{r_1^{4n+p} - r_2^{4n+p}}{r_1 - r_2} \right) - \left(\frac{r_2^{2n} r_1^{2n+p} - r_1^{2n} r_2^{2n+p}}{r_1 - r_2} \right) \tag{8}$$
$$= F_{k,4n+p} - \frac{1}{r_1 - r_2} \left[(-1)^{2n} r_1^p - (-1)^{2n} r_2^p \right]$$
$$= F_{k,4n+p} - F_{k,p}$$

Proposition

$$F_{k,4n+p} + F_{k,p} = F_{k,2n+p}L_{k,2n}, \text{ where } n \geq 1 \text{ and } p \geq 0.$$

Proof

$$\begin{aligned}
F_{k,2n+p}L_{k,2n} &= \left(\frac{r_1^{2n+p} - r_2^{2n+p}}{r_1 - r_2}\right)\left(r_1^{2n} + r_2^{2n}\right) \\
&= \left(\frac{r_1^{4n+p} - r_2^{4n+p}}{r_1 - r_2}\right) + \left(\frac{r_2^{2n} r_1^{2n+p} - r_1^{2n} r_2^{2n+p}}{r_1 - r_2}\right) \\
&= F_{k,4n+p} + \frac{1}{r_1 - r_2}\left[(-1)^{2n} r_1^p - (-1)^{2n} r_2^p\right] \\
&= F_{k,4n+p} + F_{k,p}
\end{aligned} \tag{9}$$

Proposition

$$L_{k,4n+p} - L_{k,p} = (r_1 - r_2)^2 F_{k,2n}F_{k,2n+p}, \text{ where } n \geq 1 \text{ and } p \geq 0.$$

Proof

$$\begin{aligned}
&(r_1 - r_2)^2 F_{k,2n}F_{k,2n+p} \\
&= (r_1 - r_2)^2\left(\frac{r_1^{2n} - r_2^{2n}}{r_1 - r_2}\right)\left(\frac{r_1^{2n+p} - r_2^{2n+p}}{r_1 - r_2}\right) \\
&= (r_1^{4n+p} + r_2^{4n+p}) - (r_1 r_2)^{2n}(r_1^p + r_2^p) \\
&= L_{k,4n+p} - L_{k,p}
\end{aligned} \tag{10}$$

Proposition

$$L_{k,4n+p} + L_{k,p} = L_{k,2n}L_{k,2n+p}, \text{ where } n \geq 1 \text{ and } p \geq 0.$$

Proof

$$\begin{aligned} L_{k,2n}L_{k,2n+p} &= (r_1^{2n} + r_2^{2n})(r_1^{2n+p} + r_2^{2n+p}) \\ &= (r_1^{4n+p} + r_2^{4n+p}) + (r_1 r_2)^{2n}(r_1^p + r_2^p) \\ &= L_{k,4n+p} + L_{k,p} \end{aligned} \tag{11}$$

Proposition

$$F_{k.4n+p} - F_{k,p} = F_{k,n}L_{k,n}L_{k,2n+p}, \text{where } n \geq 1 \text{and } p \geq 0.$$

Proof

$$\begin{aligned} F_{k,n}L_{k,n}L_{k,2n+p} &= \left(\frac{r_1^n - r_2^n}{r_1 - r_2}\right)\left(r_1^n + r_2^n\right)\left(r_1^{2n+p} + r_2^{2n+p}\right) \\ &= \left(\frac{r_1^n - r_2^n}{r_1 - r_2}\right)\left(r_1^{3n+p} + r_2^{3n+p} + r_1^n r_2^{2n+p} + r_2^n r_1^{2n+p}\right) \\ &= \left(\frac{r_1^{4n+p} - r_2^{4n+p}}{r_1 - r_2}\right) - \left(\frac{r_1^p - r_2^p}{r_1 - r_2}\right) \\ &= F_{k.4n+p} - F_{k,p} \end{aligned} \tag{12}$$

Proposition

$$L_{k.4n+p} - L_{k,p} = (r_1 - r_2)^2 F_{k,n}L_{k,n}F_{k,2n+p},$$

$\text{where } n \geq 1 \text{and } p \geq 0.$

Proof

$$\begin{aligned}
&(r_1 - r_2)^2 F_{k,n} L_{k,n} F_{k,2n+p} \\
&= (r_1 - r_2)^2 \left(\frac{r_1^n - r_2^n}{r_1 - r_2} \right) \left(r_1^n + r_2^n \right) \left(\frac{r_1^{2n+p} - r_2^{2n+p}}{r_1 - r_2} \right) \\
&= \left(r_1^n - r_2^n \right) \left(r_1^{3n+p} - r_2^{3n+p} - r_1^n r_2^{2n+p} + r_2^n r_1^{2n+p} \right) \\
&= \left(r_1^{4n+p} + r_2^{4n+p} \right) - \left(r_1^p + r_2^p \right) \\
&= L_{k.4n+p} - L_{k,p}
\end{aligned} \tag{13}$$

Proposition

$$F_{k.3n+p} - (-1)^n F_{k.n+p} = F_{k.n} L_{k.2n+p}, \text{where } n \geq 1 \text{ and } p \geq 0.$$

Proof

$$\begin{aligned}
&F_{k,n} L_{k,2n+p} = \left(\frac{r_1^n - r_2^n}{r_1 - r_2} \right) \left(r_1^{2n+p} + r_2^{2n+p} \right) \\
&= \left(\frac{r_1^{3n+p} - r_2^{3n+p}}{r_1 - r_2} \right) - \left(\frac{r_2^n r_1^{2n+p} - r_1^n r_2^{2n+p}}{r_1 - r_2} \right) \\
&= F_{k,3n+p} - \frac{(-1)^n}{r_1 - r_2} \left(r_1^{n+p} - r_2^{n+P} \right) \\
&= F_{k.3n+p} - (-1)^n F_{k,n+p}
\end{aligned} \tag{14}$$

In [13], if we take $i = n \text{ and } n = n + p$ in Lemma 1(i), then we have the following proposition.

Proposition

$$F_{k.3n+p} + (-1)^n F_{k,n+p} = F_{k,2n+p} L_{k,n}, \text{where } n \geq 1 \text{ and } p \geq 0.$$

Proof

$$F_{k,2n+p}L_{k,n} = \left(\frac{r_1^{2n+P} - r_2^{2n+p}}{r_1 - r_2}\right)\left(r_1^n + r_2^n\right)$$

$$= \left(\frac{r_1^{3n+p} - r_2^{3n+p}}{r_1 - r_2}\right) + \left(\frac{r_1^{2n+p}r_2^n - r_2^{2n+p}r_1^n}{r_1 - r_2}\right) \quad (15)$$

$$= F_{k,3n+p} + \frac{(-1)^n}{r_1 - r_2}\left(r_1^{n+p} - r_2^{n+P}\right)$$

$$= F_{k.3n+p} + (-1)^n F_{k,n+p}$$

Proposition

$$L_{k.3n+p} - (-1)^n L_{k,n+p} = (r_1 - r_2)^2 F_{k,n}F_{k,2n+p},$$

where $n \geq 1$ *and* $p \geq 0$.

Proof

$$(r_1 - r_2)^2 F_{k,n}F_{k,2n+p}$$

$$= (r_1 - r_2)^2\left(\frac{r_1^n - r_2^n}{r_1 - r_2}\right)\left(\frac{r_1^{2n+p} - r_2^{2n+p}}{r_1 - r_2}\right)$$

$$= (r_1^{3n+p} + r_2^{3n+p}) - (r_2^n r_1^{2n+p} + r_1^n r_2^{2n+p}) \quad (16) \qquad (16)$$

$$= L_{k,3n+p} - (-1)^n\left(r_1^{n+p} + r_2^{n+P}\right)$$

$$= L_{k,3n+p} - (-1)^n L_{k,n+p}$$

In,[13] if we take $i = n and\ ni = 2n + p$ in Lemma 1(ii), then we have the following proposition.

Proposition

$L_{k.3n+p} + (-1)^n L_{k,n+p} = L_{k,n}L_{k,2n+p}$, *where* $n \geq 1$ *and* $p \geq 0$.

Proof

$$\begin{aligned}
L_{k,n}L_{k,2n+p} &= \left(r_1^n + r_2^n\right)\left(r_1^{2n+p} + r_2^{2n+p}\right) \\
&= (r_1^{3n+p} + r_2^{3n+p}) + (r_2^n r_1^{2n+p} + r_1^n r_2^{2n+p}) \\
&= L_{k,3n+p} + (-1)^n \left(r_1^{n+p} + r_2^{n+P}\right) \\
&= L_{k,3n+p} + (-1)^n L_{k,n+p}
\end{aligned} \tag{17}$$

For different values of p, Proposition (3.1) to Proposition (3.10) can be expressed for even and odd k-Fibonacci and k-Lucas numbers.

Identities (8) to (17) give a relation between k-Fibonacci and k-Lucas numbers by using their subscripts.

CONCLUSION

In this paper we present generalized identities involving common factors of k- Fibonacci and k-Lucas numbers. Mainly Binet's formula employ for the identities. The concept can be executed for generalized second order recursive sequences as well as polynomials.

REFERENCES

1. Benjamin, A. T. and Quinn, J. J., "Recounting Fibonacci and Lucas identies", The College Mathematics Journal, 30 (5) (1999), 359-366.
2. Falcon, S. & Plaza, A., "On the Fibonacci k-numbers", Chaos, Solitons & Fractals, 32 (5), (2007), 1615-24.
3. Falcon S. and Plaza A., The k-Fibonacci hyperbolic functions, Chaos Solitons and Fractals, 38 (2) (2008), 409-420.
4. Falcon, S., "On the k-Lucas Numbers", Int. J. Contemp. Math. Sciences, 6 (21), (2011), 1039-1050.
5. Hoggatt, V.E. Jr., "Fibonacci and Lucas Numbers", Houghton – Mifflin Co., Boston (1969).
6. Horadam A. F., A generalized Fibonacci sequences, Mathematical Magazine. 68 (1961), 455-459.
7. Horadam A. F. and Shanon A. G.., Generalized Fibonacci triples, American Mathmatical Monthly, 80 (1973), 187-190.
8. Koshy, T., "Fibonacci and Lucas Numbers with Applications", John Wiley, New York (2001).

9. Kilic E., The Binet formula, sums and representation of generalized Fibonacci p-numbers, Eur. J. Combin, 29 (3) (2008), 701-711.
10. Panwar, Y. K., B. Singh and Gupta, V. K., “identities of Common Factors of generalized Fibonacci, Jacobsthal and Jacobsthal-Lucas numbers”, Applied Mathematics and Physics, 1 (4), (2013), 126-128.
11. Thongmoon, M., “New Identities for the Even and Odd Fibonacci and Lucas Numbers”, Int. J. Contemp. Math. Sciences, 4 (14), (2009), 671-676.
12. Thongmoon, M., “Identities for the common factors of Fibonacci and Lucas numbers”. International Mathematical Forum, 4 (7), (2009), 303-308.
13. Yilmaz, N., Taskara, N., Uslu,K., Yazlik, Y., On the binomial sums of k-Fibonacci and k-Lucas Sequences, American institue of Physics (AIP) Conf. Proc.

14

EXACT SOLITON SOLUTIONS OF THE 1D GENERALIZED GROSS-PITAEVSKII EQUATION WITH QUADRATIC POTENTIAL AND PARAMETERIZED NONLINEARITY

Ying Wang[1], Qing-Chun Zhou[1], Ya-Lun Li[1], Yu Zhou[1,]

[1] School of Mathematics and Physics, Jiangsu University of Science and Technology, Jiangsu, China

ABSTRACT

We investigate the 1D generalized Gross-Pitaevskii equation (GGPE) with quadratic potential and parameterized nonlinearity. The coefficients of terms of GGPE studied are arbitrary functions of time t. The exact solution(s) of the GGPE are obtained via expansion method with particular soliton features highlighted.

INTRODUCTION

It is well known that in reality many systems behave nonlinearly and described by nonlinear partial differential equations. Among these is the nonlinear Schrodinger equation that has been applied extensively and play principal role in modeling so many phenomena like the functioning of optical fibre and the realization of Bose-Einstein condensation (BEC). Due to the particular feature that the nonlinear interaction could be tuned flexibly via Feshbach resonance, BEC attracts a lot of attention in the study of nonlinear dynamics like controllable bright/dark soliton features and Gross-Pitaevskii equation (GPE) has been successfully applied to this particular application scenario and well investigated [1-8][1].

But in order to fully incorporate the mechanism of tunable Feshbach resonance, the generalized Gross-Pitaevskii equation (GGPE) with nonlinearity of arbitrary power index of modulus of wave function is more appropriate [9, 10]. The investigation of such GGPE with parameterized nonlinearity however is relatively rare. In this paper, we will investigate the model for the system of BEC with tunable Feshbach resonance control in the cigar-shaped harmonic trap and gravitational field that is described by the 1D GGPE. To give a more thorough and general treatment, we allow the coefficients of external(quadratic) potential and nonlinear terms to be arbitrary functions of t. We demonstrate that such 1D GGPE with parameterized nonlinearity and quadratic external potential is analytically solvable and obtain its exact solution through modified $(1/G, G'/G)$ expansion method [11].

The paper is arranged as follows, in the next section, we give the outline of solution finding strategies and main results, followed by the commentary explanation and discussion, with the concluding and summarization statement for the work presented given in the last section.

Methodologies

Problem Formulation

The dimensionless 1D generalized Gross-Pitaevskii equation (GGPE) to be studied reads

$$\begin{aligned} i\frac{\partial\psi(x,t)}{\partial t} &= -\frac{\partial^2}{\partial x^2}\psi(x,t)+k_2(t)x^2\psi(x,t) \\ &+k_1(t)x\psi(x,t)+g(t)\,|\,\psi(x,t)\,|^{2\gamma}\,\psi(x,t) \end{aligned} \tag{1}$$

Where γ is a real constant power index in the nonlinear term determined by the experiment that is discussed in Refs. [12-17] The first term on the right hand side of (1) is dispersion term, the second term on the RHS is arising from external harmonic magnetic trapping potential, the third term is from gravitational potential, the fourth term is the interaction term with Landau coefficient $g(t)>0$ corresponding to the repulsive interaction, $g(t)<0$ corresponding to the attractive interaction. The GGPE (1) is the reduction from its 3D analog in certain scenario like cigar-shaped elongated external harmonic potential $V(r)=\frac{1}{2}m(\omega_\rho^2\rho^2+\omega_x^2x^2)$ with $\omega_\rho >> \omega_x$, similar as what were discussed in some prior work[18, 19], $k_2(t), k_1(t)$ are chosen to be arbitrary function of time t.

To eliminate the interdependence between $k(t)$ and $g(t)$ to make the 1D GGPE analytically solvable, we introduce a parameter function $\sigma(t)$ into the coefficients of the above equation through the following transformation

$$x' = \sigma(t')x+\sigma_2(t'), \tag{2}$$

$$t' = t, \tag{3}$$

$$\psi(x,t) = \sigma^{1/2}(t') \times \exp[i(\frac{1}{2}\frac{\sigma_{t'}(t')}{\sigma(t')}x^2 + \sigma_3(t)x + \sigma_4(t))]\phi(x',t') \tag{4}$$

Formula (2), (3) and (4) can be thought as modified lens-type transformation[20, 21]. We could express $\sigma_2(t), \sigma_3(t)$, $\sigma_4(t)$ using $\sigma(t)$ with the following three constraint formula

$$\sigma_2(t) = \frac{k_1(t) - [\sigma_{2'}(t)/\sigma_2(t)]'}{2\sigma(t)[k_2(t)/\sigma^2(t) - 1/4\sigma_t^2(t)/\sigma^2(t) - 1/4(\sigma_t(t)/\sigma(t))']} \tag{5}$$

$$\sigma_3(t) = -\sigma_{2'}(t)/2\sigma(t) \tag{6}$$

$$\sigma_{4'}(t) = -1/4\sigma_{2'}^2(t)/\sigma^2(t) + \frac{[k_1(t) - 1/2(\sigma_{2'}(t)/\sigma(t))']^2}{4\sigma(t)[k_2(t)/\sigma^2(t) - 1/4\sigma_t^2(t)/\sigma^2(t) - 1/4(\sigma_t(t)/\sigma(t))']} \tag{7}$$

with $\sigma(t)$ to be determined later. Substituting (2), (3) and (4) into Eq.(1) and changing notation from (x',t') to (x,t) we get the generalized GP equation with modified coefficients as

Assume wavefunction is of the form

$$i\phi_t + \sigma^2(t)\phi_{xx} + [\frac{k(t)}{\sigma^2(t)} - (\frac{\sigma_t(t)}{2\sigma(t)})^2 - \frac{1}{4}(\frac{\sigma_t(t)}{\sigma(t)})_t]x^2\phi + g(t)\sigma^\gamma(t)|\phi|^{2\gamma}\varphi = 0 \tag{8}$$

substitute (9) into Eq.(8), we get the following equations for $v(x,t)$ and $\theta(x,t)$

$$\phi(x,t) = v^{1/\gamma}(x,t)e^{i\theta(x,t)} \tag{9}$$

$$v^2\theta_t + \sigma^2(t)(a_0 v v_{xx} + b_0 v_x^2 + v^2\theta_x^2) + \alpha(t)x^2v^2 + \beta(t)v^4 = 0 \tag{10}$$

$$v_t + \sigma^2(t)(2v_x\theta_x + \gamma v\theta_{xx}) = 0 \tag{11}$$

where $\alpha(t) = k(t)/\sigma^2(t) - 1/4(\sigma_t(t)/\sigma(t))^2 - (\sigma_t(t)/4\sigma(t))_t$, $\beta(t) = g(t)\sigma^\gamma(t)/\hbar\omega$, $a_0 = -1/\gamma$,and $b_0 = -(1-\gamma)/\gamma$ are constants. In the ensuing equations solving steps we will work on Eqs.(10-11) for the final solution of the GGPE.

Method Outline

In order to obtain the exact analytical solution of Eqs.(2.1), we consider the (G'/G ,$1/G$)-expansion method, which is applicable to solving the nonlinear partial differential equation of the form

$$F(u, u_t, u_x, u_{xx}, \ldots) = 0 \tag{12}$$

Where F is the polynomial of unknown function $u(x,t)$ and its partial derivatives of various order. The basic idea of (G'/G ,$1/G$)-expansion is trying to express the unknown function as polynomial of $G'(\xi)/G(\xi)$ and $1/G$, with $G(\xi)$ defined as function of $\xi = p(t)x + q(t)$ like

$$\frac{d^2}{d\xi^2}G(\xi) + \lambda G(\xi) = 0 \tag{13}$$

where λ is constant. There are three cases for the solution of Eq.(13):

Case 1 When $\lambda < 0$,

$$G(\xi) = A_1 \sinh\sqrt{-\lambda}\xi + A_2 \cosh\sqrt{-\lambda}\xi, \tag{14}$$

and we have

$$1/G^2 = -\frac{1}{\lambda\rho}[(G'/G)^2 + \lambda] \tag{15}$$

where A_1 and A_2 are two arbitrary constants and $\rho = A_1^2 - A_2^2$.

Case 2 When $\lambda > 0$,

$$G(\xi) = A_1 \sin\sqrt{\lambda}\xi + A_2 \cos\sqrt{\lambda}\xi, \tag{16}$$

and we have

$$1/G^2 = \frac{1}{\lambda\rho}[(G'/G)^2 + \lambda] \tag{17}$$

where A_1 and A_2 are two arbitrary constants and $\rho = A_1^2 + A_2^2$.

Case 3 When $\lambda = 0$, the general solution of Eq.(13) is

$$G(\xi) = A_1\xi + A_2, \tag{18}$$

and we have

$$1/G^2 = \frac{1}{A_1^2}[(G'/G)^2] \tag{19}$$

where A_1 and A_2 are two arbitrary constants. From (15,17,19), we can express the the solution $u(x,t)$ of (12) as

$$u(x,t) = \sum_{i=0}^{m} z_i(t)(\frac{G'(\xi)}{G\xi})^i + \sum_{i=1}^{m} h_i(t)\frac{1}{G(\xi)}(\frac{G'(\xi)}{G\xi})^{i-1},$$
$$h_m^2(t) + z_m^2(t) \neq 0 \tag{20}$$

Substituting (20) into the original nonlinear partial differential equation (12) and making use of relations (13) and (15,17,19), m can be set by balancing between the highest differential term and nonlinear term and express F in (12) as a polynomial of $G'(\xi)/G(\xi)$ plus another polynomial of $G'(\xi)/G(\xi)$ times $1/G(\xi)$. The Eq. (12) is solved by setting the coefficients of all terms $(G'^i(\xi)/G^i(\xi)$ and $G'^j(\xi)/G^j(\xi)1/G(\xi))$ of F to zero. This will result in a set of over determined ODEs for $h_i(t)$ and $z_j(t)$, that will put the unknown function $u(x,t)$ shown in (20) in a explicit form if the ODEs are solved consistently.

Procedure Details and Results

To proceed with the $(G'/G, 1/G)$-expansion problem solving strategy for Eq.(2.1), we assume $v(x,t)$ and $\theta(x,t)$ in Eq.(2.1) are of the form

$$v(x,t)=\sum_{i=0}^{m} z_i(t)\left(\frac{G'(\xi)}{G\xi}\right)^i+\sum_{i=1}^{m} h_i(t)\frac{1}{G(\xi)}\left(\frac{G'(\xi)}{G\xi}\right)^{i-1}, \quad h_m^2(t)+z_m^2(t)\neq 0 \tag{21}$$

$$\theta(x,t)=\Phi(t)x^2+\Gamma(t)x+\Omega(t) \tag{22}$$

with ξ and $G(\xi)$ defined as that in (13). The highest power index m in (21) is set by balancing between the highest derivative term and the nonlinear term when substituting (21) and (22) into Eq.(2.1) and making use of (13) and (15,17,19). The balancing formula is $2m+2=4m$, so $m=1$; hence $v(x,t)$ in (21) can be written in two possible formats as

$$v(x,t)=h(t)\frac{1}{G(\xi)}+f(t) \tag{23}$$

$$v(x,t)=z(t)\frac{G'(\xi)}{G(\xi)}+f(t) \tag{24}$$

The problem solving details for the above two cases are very similar, we show the explicit details for (23). After plugging the formatted solution (21) and (22) into Eq.(10-11), making use of formula (13) and (15,17, 19), we have the equations of (10-11) expressed as polynomials for $x^j(G'(\xi)/G(\xi))^i(1/G(\xi))^r$, where $j=0,1,2$; $i\in[0,4]$, and $r=0$ or 1. Setting the coefficients of every term $x^j(G'(\xi)/G(\xi))^i(1/G(\xi))^r$ to zero, we get the following set of ODEs, (define M$=G'(\xi)/G(\xi)$,N$=1/G(\xi)$).

For Eq.(10), we have

$$x^2\mathrm{M}^2: \quad h^2(t)[\Phi'(t)+4\sigma^2(t)\Phi^2(t)+\alpha(t)]=0, \tag{25}$$

$$x^2N: \quad 2h(t)f(t)[\Phi'(t)+4\sigma^2(t)\Phi^2(t)+\alpha(t)]=0, \tag{26}$$

$$x^2 : (f^2(t) - h^2(t)/\rho)[\Phi'(t) + 4\sigma^2(t)\Phi^2(t) + \alpha(t)] = 0 \tag{27}$$

$$xM^2 : \quad h^2(t)[\Gamma'(t) + 4\sigma^2(t)\Phi(t)\Gamma(t)] = 0, \tag{28}$$

$$xN : \quad 2h(t)f(t)[\Gamma'(t) + 4\sigma^2(t)\Phi(t)\Gamma(t)] = 0, \tag{29}$$

$$x : (f^2(t) - h^2(t)/\rho)[\Gamma'(t) + 4\sigma^2(t)\Phi(t)\Gamma(t)] = 0 \tag{30}$$

and

$$\begin{aligned} &M^4 : h^2(t)[2a_0\sigma^2(t)p^2(t) \\ &+b_0\sigma^2(t)p^2(t) - \beta(t)h^2(t)/(\lambda\rho)] = 0 \end{aligned} \tag{31}$$

$$M^2N : \quad \beta(t)f(t)h^3(t)/(\lambda\rho) = 0, \tag{32}$$

$$\begin{aligned} &M^2 : h^2(t)[\Omega'(t) + \sigma^2(t)\Gamma(t)^2 \\ &+\sigma^2(t)p^2(t)(3a_0 + b_0) + 6\beta(t)f^2(t)] = 0, \end{aligned} \tag{33}$$

$$\begin{aligned} &N : f(t)h(t)[\Omega'(t) + \sigma^2(t)\Gamma(t)^2 \\ &+\beta(t)(f^2(t) - h^2(t)/\rho)] = 0, \end{aligned} \tag{34}$$

$$\begin{aligned} &M^0 : \quad (f^2(t) - h^2(t)/\rho)(\Omega'(t) + \sigma^2(t)\Gamma(t)^2) \\ &+ a_0\frac{\lambda}{\rho}\sigma^2(t)h^2(t)p^2(t) \\ &+ \beta(t)[(f^2(t) - h^2(t)/\rho)^2 - 4h^2(t)f^2(t)/\rho] = 0, \end{aligned} \tag{35}$$

While for Eq.(11) we have

$$xMN : h(t)[p'(t) + 4\sigma^2(t)\Phi(t)p(t)] = 0 \tag{36}$$

and

$$MN:\ h(t)[q'(t)+4\sigma^2(t)p(t)\Gamma(t)]=0, \tag{37}$$

$$N:\ h'(t)+2\gamma\sigma^2(t)\Phi(t)h(t)=0, \tag{38}$$

$$M^0:\ f'(t)+2\gamma\sigma^2(t)\Phi(t)f(t)=0, \tag{39}$$

From the above ODEs (25)-(39), we can see by introducing new parameter function $\sigma(t)$, both $k_i(t)$ (in $\alpha(t)$) and $g(t)$ (in $\beta(t)$) can change freely.

The above ODEs can be solved analytically, with appropriate boundary conditions, and the solutions could be expressed as (We express the unknown functions of t in $\Phi(t)$ and $\sigma(t)$ whose analytical forms are analyzed in the end):

$$p(t)=C_1\exp[-4\int\sigma^2(s)\Phi(s)ds] \tag{40}$$

$$q(t)=-2C_1C_2\int[\sigma^2(s)\cdot\exp(-8\int\sigma^2(z)\Phi(z)dz)]ds \tag{41}$$

$$\Gamma(t)=C_2\exp[-4\int\sigma^2(s)\Phi(s)ds] \tag{42}$$

$$\begin{aligned}\Omega(t)=&-[(3a_0+b_0)C_1^2+C_2^2]\\&\times\int[\sigma^2(s)\cdot\exp(-8\int\sigma^2(z)\Phi(z)dz)ds]\end{aligned} \tag{43}$$

$$h(t)=C_3\exp[-2\gamma\int\sigma^2(s)\Phi(s)ds] \tag{44}$$

$$f(t)=0 \tag{45}$$

$$\begin{aligned}G(\xi)=\sqrt{-\rho}\cosh(\sqrt{-\lambda}(\xi-\xi_0)),\\ \lambda<0, g(t)<0\end{aligned} \tag{46}$$

$$G(\xi)=\sqrt{\rho}(\xi-\xi_0),\lambda=0,g(t)<0 \tag{47}$$

$$G(\xi) = \sqrt{-\rho}\cos(\sqrt{\lambda}(\xi - \xi_C)), \lambda > 0, g(t) > 0 \tag{48}$$

$$\begin{aligned} v(x,t) &= \sqrt{-\rho}h(t)\operatorname{sech}[\sqrt{-\lambda}(p(t)x + q(t) - \xi_0)], \\ &\lambda < 0, g(t) < 0 \end{aligned} \tag{49}$$

$$v(x,t) = \sqrt{\rho}(p(t)x + q(t) - \xi_0), \lambda = 0, g(t) < 0 \tag{50}$$

$$\begin{aligned} v(x,t) &= \sqrt{-\rho}h(t)\sec[\sqrt{\lambda}(p(t)x + q(t) - \xi_0)], \\ &\lambda > 0, g(t) > 0 \end{aligned} \tag{51}$$

Where C_1, C_2, λ, ρ and ξ_0 are arbitrary constants, $\sigma(t)$ and $\Phi(t)$ are determined by the following pair of equations:

$$\sigma'(t) + \frac{g'(t)}{(2-\gamma)g(t)}\sigma(t) - 4\sigma^3(t)\Phi(t) = 0, \tag{52a}$$

$$\Phi'(t) + 4\sigma^2(t)\Phi^2(t) + \alpha(t) = 0, \tag{52b}$$

where $g(t)$ and $k_i(t)$ (in $\alpha(t)$) are arbitrary functions of t; however, we always have constants $g(t) = g$ and $k_i(t) = k_i$ in experiments. So $\sigma(t)$ can be directly expressed by $\Phi(t)$ from Eq. (52) as

$$\sigma(t) = \pm(C_4 - 8\int\Phi(s)ds)^{-1/2}$$

with specific constant C_4. Eq.(53) could also be integrated with constant k_i, g using Maple for example. Since generally $\Phi(t)$ can not be expressed in simple rudimentary function form, we omit the formula that shows explicit the expression for $\Phi(t)$ here. Fortunately, for interested case where both $k_i(t)$ and $g(t)$ are constants, we can bypass the explicit expression for $\Phi(t)$ and reach directly the explicit expressions for the parameters in final solution, as explained in the following section.

COMMENTS AND DISCUSSIONS

For the solution we obtained in previous section, returning to the original coordinates, from (2),(4),(9) and (49-51), the final solution take the following form

$$|\psi(x,t)| = \begin{cases} \sqrt{\sigma(t)}(-\lambda h^2(t))^{1/2\gamma} \\ \times \operatorname{sech}^{1/\gamma}[\sqrt{-\lambda}\chi(t) + q_0(t)], \lambda < 0, g(t) < 0 \\ \sqrt{\rho}(x - \sigma_2(t)/\sigma(t))^{1/2\gamma}, \lambda = 0, g(t) < 0 \\ \sqrt{\sigma(t)}(\lambda h^2(t))^{1/2\gamma} \\ \times \sec^{1/\gamma}[\sqrt{\lambda}\chi(t) + q_0(t)], \lambda > 0, g(t) > 0 \end{cases} \tag{54a}$$

where, $\chi(t) = p(t)\sigma(t)(x - \sigma_2(t)/\sigma(t)), q_0(t) = q(t) - \xi_0$, $\sigma(t), h(t), p(t), q(t)$ are given in (40-45). We are particularly interested in the solution of bright soliton type given in first case in (54a) for $\lambda < 0$ and $g(t) < 0$which corresponds to the attractive nonlinear interaction. In this case, with $k(t) = k$, $g(t) = g$ as constants, from the previous results (40-45) and (54), it is not hard to see that $p(t)\sigma(t) = \kappa$ is constant, $q(t) = \varpi t$ (ϖ is constant), the soliton's shape remain unchanged with time and the peak moves with constant speed ϖ/κ perturbed by the gravitational field shown in $\sigma_2(t)/\sigma(t)$. When $g(t)$ (or $k(t)$)depends on t, $p(t)\sigma(t)$, which indicates the compressing ratio of the travelling wave along the x-direction, generally is not constant and the format of $\sigma_2(t)/\sigma(t) - q(t)/(p(t)\sigma(t))$ is complicated, which mean that there is shape variation in the traveling wave and the wave peak move with acceleration.

Also our analysis could be extended to the case of repulsive nonlinear interaction ($g(t) > 0$) and $\lambda < 0$ which corresponds to the solution format (24). In a similar manner as (23), we can obtain a very similar set of ODEs as that of (25-39), and reach an analytical solution of dark soliton type for the GGPE as follow

$$|\psi(x,t)| = \sqrt{\sigma(t)}(-\lambda h^2(t))^{1/2\gamma} \tanh^{1/\gamma}[\sqrt{-\lambda}\chi(t) + q_0(t)] \tag{54b}$$

with $\sigma(t), h(t), p(t), q(t)$ given by the same formula as (40-45). When $k(t)$ and $g(t)$ are constants, $p(t)\sigma(t)$is constant, the dark soliton shape does not change, also when $g(t)$ or $k(t)$ (or both) has time dependence, generally there will be modulation effects on the shape of traveling wave which normally has non-zero acceleration.

The analysis and results presented to this point should be typical, we could leave for future work the extended study, which include for example more mathematical sound solution format(s) based on expansion method.

CONCLUSION

In this paper, we find exact solutions for a much broader category of the generalized 1D Gross-Pitaevskii equation with general quadratic potential. The coefficient functions of the terms of the generalized GPE are arbitrary. We also give a generalized treatment of the nonlinearity with parameterized power index. We eliminate the restriction shown in some previous work on GPE which require a constraint formula connecting variable coefficients of the equation terms. We reach the exact analytical solutions of the GGPE using modified $(G'/G, 1/G)$ expansion method, and for typical setting for the parameters, we get soliton like traveling wave solutions which are similar to the solutions obtained in previous work on GPE but with general time modulation to the amplitude and shape of the traveling wave arising from the variable coefficients of the equation terms.

ACKNOWLEDGEMENT

This work was supported by NSF-China under Grant Nos. 11105039, 11204274, and 11205071.

REFERENCES

1. Mendonça, J. T. and Terças, H Physics of Ultra-Cold Matter, Springer, New York, 2013.
2. Levin, K., Fetter, A. and Stamper-Kurn, D. Ultracold Bosonic and Fermionic Gases, Elsevier, Oxford, 2012.
3. Zhang, A. X. and Xue, J. K., "Dynamics of Bright/Dark Solitons in Bose-Einstein Condensates with Time-Dependent Scattering Length and External Potential," Chin. Phys. Lett. 25. 39. 2008.
4. Liu, H., He, D., Lou, S. Y., and He, X. T., "Bright Soliton Solutions in Degenerate Fermi Gas near Feshbach Resonance," Chin. Phys. Lett. 26.120308. 2009.
5. Zhang, S., Ba, J. M., Sun, Y. N. and Ling, D. Z., "A Generalized (G'/G)-Expansion Method for the Nonlinear Schrodinger Equation with Variable Coefficients,"Naturforsch. 64a 691. 2009.
6. Fei, J. X. and Zheng, C. L., "Exact Projective Excitations of a Generalized (3+1)-Dimensional Gross-Pitaevskii System with Varying Parameters," Chin. J. Phys. 51. 200. 2013.
7. Trallero-Ginera, C., Drake-Pereza, J. C., Lopez-Richardb, V. and Birmanc, J. L., "Formal analytical solutions for the Gross–Pitaevskii equation," Physica D 237. 2342. 2008.

8. Atre, R,, Panigrahi, P. K. and Agarwal, G. S., "Class of solitary wave solutions of the one-dimensional Gross-Pitaevskii equation," Phys. Rev. E 73. 056611. 2006.

9. Feshbach, H., "Unified theory of nuclear reactions," Ann. Phys. (N.Y.), 5. 357. 1958.; ibid. 19. 287. 1962.; Inouye, S., Andrews, M. R., Stenger, J., Miesner, H., Stamper-Kurn, D. M. and Ketterle, W., "Observation of Feshbach resonances in a Bose–Einstein condensate",Nature 392. 151. 1998.

10. Zwerger W The BCS-BEC Crossover and the Unitary Fermi Gases, Springer, Berlin, 2012.

11. Li, L. X., LI, E. Q. and Wang, M. L., " The (G'/G, 1/G)-expansion method and its application to travelling wave solutions of the Zakharov equations," Appl. Math. J. Chinese Univ. 25 (4). 454. 2010.

12. Heiselberg, H., "Collective Modes of Trapped Gases at the BEC-BCS Crossover," Phys. Rev. Lett. 93. 040402. 2004.

13. Yin, J., Ma, Y. and Huang, G., "Analytical expressions of collective excitations for trapped superfluid Fermi gases in a BCS-BEC crossover," Phys. Rev. A 74. 013609. 2006.

14. Zhou, Y. and Huang. G., "Collective modes of low-dimensional superfluid Fermi gases in the BCS-BEC crossover: Time-dependent variational analysis," Phys. Rev. A 75. 023611. 2007.

15. Kim, Y. E. and Zubarev, A. L., "Time-dependent density-functional theory for trapped strongly interacting fermionic atoms," Phys. Rev. A 70. 033612. 2004.

16. Ghosh, T. K. and Machida, K., "Sound velocity and multibranch Bogoliubov spectrum of an elongated Fermi superfluid in the BEC-BCS crossover," Phys. Rev. A 73. 013613. 2006.

17. Hu, H., Minguzzi, A., Liu, X. J. and Tosi, M. P., "Collective Modes and Ballistic Expansion of a Fermi Gas in the BCS-BEC Crossover," Phys. Rev. Lett. 93. 190403. 2004.

18. Trallero-Giner, C., Cipolatti, R. and Liew Timothy, C. H., "One-dimensional cubic-quintic Gross-Pitaevskii equation for Bose-Einstein condensates in a trap potential," Eur. Phys. J. D 67. 143. 2013.

19. Muryshev, A., Shlyapnikov, G. V., Ertmer, W., Sengstock, K. and Lewenstein, M., "Dynamics of Dark Solitons in Elongated Bose-Einstein Condensates," Phys. Rev. Lett. 89. 110401. 2002.

20. Theocharis, G. et al, "Modulational instability of Gross-Pitaevskii-type equations in 1+1 dimensions,"Phys. Rev. A 67. 063610. 2003.

21. Sulem, C. and Sulem, P. L., The Nonlinear Schrödinger Equation, Springer, New York, 1999.

15

ON LOCATION OF ZEROS OF POLYNOMIALS

M S Pukhta*

[1] Division of Agri. Statistics, Sher-e-Kashmir, University of Agricultural Sciences and Technology of Kashmir, India

ABSTRACT

If $p(z)=\sum_{j=0}^{n}\alpha_j z^j$ be a polynomial of degree n such that $a_j=a_j+ib_j$ where a_j and b_j, j = 0, 1,n are real numbers. In this paper we obtain a generalization of well known result of Eneström -Kakeya concerning the bounds for the moduli of the zeros of polynomials with complex coefficients which improve upon some results due to A. Aziz and Q.G Mohammad and others.

INTRODUCTION AND STATEMENT OF RESULTS

The following result known as Eneström-Kakeya theorem, is well known in the theory of distribution of zeros of polynomials was firstly proved by Eneström[2] and Kakeya[5] and later independently by Hurwitz[3].

Theorem 1.1. If $p(z)=\sum_{j=0}^{n}a_j z^j$ is a polynomial of degree n such that

$$a_n \geq a_{n-1} \geq \ldots \geq a_1 \geq a_0 > 0,$$

then all the zeros of $p(z)$ lie in $|z|\leq 1$.

Joyal, Labelle and Rahman [4] extended Theorem A to the polynomial whose coefficients are monotonic but not necessarily non-negative by proving the following:

Theorem 1.2. Let $p(z)=\sum_{j=0}^{n} a_j z^j$ be a polynomial of degree n such that

$$a_n \geq a_{n-1} \geq \ldots \geq a_1 \geq a_0,$$

then all the zeros of $p(z)$ lie in

$$|z| \leq \frac{1}{|a_n|}\{a_n - a_0 + |a_0|\}.$$

Aziz and Zargar [6] relaxed the hypothesis of Eneström-Kakeya theorem and proved the following extension of Theorem 1.2.

Theorem 1.3. Let $p(z)=\sum_{j=0}^{n} a_j z^j$ be a polynomial of degree n such that for some $k \geq 1$,

$$ka_n \geq a_{n-1} \geq \ldots \geq a_1 \geq a_0,$$

then all the zeros of $p(z)$ lie in

$$|z+k-1| \leq \frac{ka_n - a_0 + |a_0|}{|a_n|}.$$

By using Schwartz lemma, Aziz and Mohammad [1] generalized Eneström -Kakeya theorem in a different way and proved:

Theorem 1.4. If $p(z)=\sum_{j=0}^{n} a_j z^j$ be a polynomial of degree n with real positive coefficients. If $t_1 > t_2 \geq 0$ can be found such that

$$\begin{aligned} &a_r t_1 t_2 + a_{r-1}(t_1 - t_2) - a_{r-2} \geq 0, \\ &r = 1,2,\ldots,n+1; a_{-1} = a_{n+1} = 0 \end{aligned},$$

then all the zeros of $p(z)$ lie in $|z| \leq t_1$.

In this paper, we prove some generalizations and extensions of Theorem 1.1, Theorem 1.2, Theorem 1.3 and Theorem 1.4. In this direction we first present the following interesting result which is generalization of Theorem 1.4.

Theorem 1.5. If $p(z)=\sum_{j=0}^{n}\alpha_j z^j$ be a polynomial of degree n such that $\alpha_j=a_j+ib_j$ where a_j and b_j, j = 0, 1, ….., n are real numbers and for certain non-negative real numbers t_1,t_2 with $t_1>t_2$ and $t_1\neq 0$

$$a_r t_1 t_2 + a_{r-1}(t_1-t_2)-a_{r-2}\geq 0, r=1,2,....,n+1$$
$$b_r t_1 t_2 + b_{r-1}(t_1-t_2)-b_{r-2}\geq 0, \quad r=1,2,....,n+1$$
$$a_{n+1}=0=a_{-1}=a_{-2}$$
$$b_{n+1}=0=b_{-1}=b_{-2}$$

then all the zeros of $p(z)$ lie in

$$|z|\leq\frac{|\alpha_n+M|t_1}{|\alpha_n|},$$

where

$$M=(|a_0|-a_0)\frac{t_2}{t_1^{n+1}}+(|b_0|-b_0)\frac{t_2}{t_1^{n+1}}.$$

Remark 1.1. If in Theorem 1.5, we assume that all the coefficients are real and positive, then M=0 and it reduces to Theorem 1.4 due to Aziz and Mohammad **[1].**

In particular, if we choose t2=0 and $t_1\equiv t$ in Theorem 1.5, then we have the following result.

Corollary 1.1. If $p(z)=\sum_{j=0}^{n}\alpha_j z^j$ be a polynomial of degree n such that $\alpha_j=a_j+ib_j$ where a_j and, b_j, j = 0, 1, 2 ……, n are real numbers. If $t>0$ be such that

$$a_n t^n \geq a_{n-1}t^{n-1}\geq ....\geq a_1 t\geq a_0$$
$$b_n t^n \geq b_{n-1}t^{n-1}\geq ....\geq b_1 t\geq b_0$$

then all the zeros of $p(z)$ lie in

$$|z|\leq\frac{|\alpha_n+M'|t}{|\alpha_n|},$$

where

$$M' = \frac{1}{t^n}\left[\left(|a_0| - a_0\right) + \left(|b_0| - b_0\right)\right].$$

Remark 1.2. If we consider all the coefficients in Corollary 1.1 to be real and choose t = 1, then we get Theorem 1.2 due to Joyal, Labelle and Rahman [4] In addition, if we choose all coefficients to be positive, then we obtain the well-known Eneström-Kakeya Theorem (Theorem 1.1).

Next we prove the following result which is also a generalization of Theorem

Theorem.1.6. Let $p(z) = \sum_{j=0}^{n} \alpha_j z^j$ be a polynomial of degree n such that $\alpha_j = a_j + ib_j$ where a_j and b_j, j = 0, 1, ….., n are real numbers and for certain non-negative real numbers t_1, t_2 with $t_1 > t_2$ and $t_1 \neq 0$

$$a_r t_1 t_2 + a_{r-1}(t_1 - t_2) - a_{r-2} \geq 0,$$
$$b_r t_1 t_2 + b_{r-1}(t_1 - t_2) - b_{r-2} \geq 0, \quad r = 2, \ldots, n$$

and for some $k \geq 1$

and

$$ka_n(t_1 - t_2) - a_{n-1} \geq 0$$
$$kb_n(t_1 - t_2) - b_{n-1} \geq 0$$
$$a_{-1} = a_{n+1} = 0 = b_{n+1} = b_{-1}$$

then all the zeros of $p(z)$ lie in

$$\left|z + (k-1)(t_1 - t_2)\right| \leq R_1,$$

where

$$R_1 = \frac{1}{|\alpha_n|}\left[\left\{ka_n(t_1 - t_2) + a_n t_2 - \frac{a_1 t_2}{t_1^{n-1}} - \frac{a_0}{t_1^{n-1}}\right\}\right.$$
$$+\left\{kb_n(t_1 - t_2) + b_n t_2 - \frac{b_1 t_2}{t_1^{n-1}} - \frac{b_0}{t_1^{n-1}}\right\}$$
$$+\left|a_1 t_1 t_2 + a_0(t_1 - t_2)\right|\frac{1}{t_1^n}$$
$$\left.+\left|b_1 t_1 t_2 + b_0(t_1 - t_2)\right|\frac{1}{t_1^n} + (|a_0| + |b_0|)\frac{t_2}{t_1^n}\right].$$

Remark 1.3. If we choose $t_2 = 0$ and $t_1 \equiv t$ in Theorem 1.6, we get the following result.

Corollary 1.2. Let $p(z) = \sum_{j=0}^{n} a_j z^j$ be a polynomial of degree n such that $\alpha_j = a_j + ib_j$ where a_j and b_j, j = 0, 1, 2,….., n are real numbers and for certain non-negative real number t

$$a_{n-1}t^{n-1} \geq a_{n-2}t^{n-2} \geq \ldots\ldots \geq a_1 t > a_0$$
$$b_{n-1}t^{n-1} \geq b_{n-2}t^{n-2} \geq \ldots\ldots \geq b_1 t > b_0$$

and for some $k \geq 1$,

$$ka_n t^n \geq a_{n-1}t^{n-1}$$
$$kb_n t^n \geq b_{n-1}t^{n-1}.$$

Then all the zeros of $p(z)$ lie in

$$|z + (k-1)t| \leq R_2,$$

where $R_2 = \frac{1}{|\alpha_n|}\left[\begin{matrix} k(a_n + b_n)t + (|a_0| + |b_0|)\frac{1}{t^{n-1}} \\ -(a_0 + b_0)\frac{1}{t^{n-1}} \end{matrix}\right].$

Remark 1.4. If we consider all the coefficients in Corollary 1.2 to be real and positive, then we obtain Theorem 1.3 due to Aziz and Zargar [1].

Remark 1.5. Again, if we choose all the coefficients in Corollary 1 to be real and consider k=1 and t=1, we get Theorem B due to Joyal, Labelle and Rahman [5].

PROOF OF THE THEOREM

Proof of Theorem 1.5. Consider the polynomial

$$\begin{aligned}
F(z) &= (t_2+z)(t_1-z)p(z) \\
&= \left\{t_1t_2+(t_1-t_2)z-z^2\right\}\left\{\begin{matrix}\alpha_n z^n+\alpha_{n-1}z^{n-1} \\ +\ldots\ldots+\alpha_1 z+\alpha_0\end{matrix}\right\} \\
&= -\alpha_n z^{n+2}+\left\{\alpha_n(t_1-t_2)-\alpha_{n-1}\right\}z^{n+1} \\
&\quad+\left\{\alpha_n t_1t_2+\alpha_{n-1}(t_1-t_2)-\alpha_{n-2}\right\}z^n+\ldots\ldots \\
&\quad+\left\{\alpha_2 t_1t_2+\alpha_1(t_1-t_2)-\alpha_0\right\}z^2 \\
&\quad+\left\{\alpha_1 t_1t_2+\alpha_0(t_1-t_2)\right\}z+\alpha_0 t_1t_2
\end{aligned}$$

$$\begin{aligned}
&= -a_n z^{n+2}+\left\{a_n(t_1-t_2)-a_{n-1}\right\}z^{n+1}+\ldots\ldots \\
&\quad+\left\{a_1t_1t_2+a_0(t_1-t_2)\right\}z+a_0t_1t_2+ \\
&\quad i\Big[-b_n z^{n+2}+\left\{b_n(t_1-t_2)-b_{n-1}\right\}z^{n+1}+\ldots\ldots \\
&\quad+\left\{b_1t_1t_2+b_0(t_1-t_2)\right\}z+b_0t_1t_2\Big].
\end{aligned}$$

Further, let

$$
\begin{aligned}
G(z) = z^{n+2}F\left(\frac{1}{z}\right) &= -\alpha_n + \{\alpha_n(t_1-t_2)-\alpha_{n-1}\}z \\
&+\{\alpha_n t_1 t_2+\alpha_{n-1}(t_1-t_2)-\alpha_{n-2}\}z^2+\ldots\ldots \\
&+\{\alpha_2 t_1 t_2+\alpha_1(t_1-t_2)-\alpha_0\}z^n \\
&+\{\alpha_1 t_1 t_2+\alpha_0(t_1-t_2)\}z^{n+1}+\alpha_0 t_1 t_2 z^{n+2} \\
&= -a_n+\{a_n(t_1-t_2)-a_{n-1}\}z \\
&+\{a_n t_1 t_2+a_{n-1}(t_1-t_2)-a_{n-2}\}z^2+\ldots\ldots \\
&+\{a_2 t_1 t_2+a_1(t_1-t_2)-a_0\}z^n \\
&+\{a_1 t_1 t_2+a_0(t_1-t_2)\}z^{n+1}+a_0 t_1 t_2 z^{n+2} \\
&+i\Big[-b_n+\{b_n(t_1-t_2)-b_{n-1}\}z \\
&+\{b_n t_1 t_2+b_{n-1}(t_1-t_2)-b_{n-2}\}z^2+\ldots\ldots \\
&+\{b_2 t_1 t_2+b_1(t_1-t_2)-b_0\}z^n \\
&+\{b_1 t_1 t_2+b_0(t_1-t_2)\}z^{n+1}+b_0 t_1 t_2 z^{n+2}\Big] \\
&= -\alpha_n+\phi(z),
\end{aligned} \tag{3.1}
$$

where

$$\phi(z)=\{a_n(t_1-t_2)-a_{n-1}\}z$$
$$+\{a_nt_1t_2+a_{n-1}(t_1-t_2)-a_{n-2}\}z^2+\ldots\ldots$$
$$+\{a_2t_1t_2+a_1(t_1-t_2)-a_0\}z^n$$
$$+\{a_1t_1t_2+a_0(t_1-t_2)\}z^{n+1}$$
$$+a_0t_1t_2z^{n+2}+i\big[\{b_n(t_1-t_2)-b_{n-1}\}z$$
$$+\{b_nt_1t_2+b_{n-1}(t_1-t_2)-b_{n-2}\}z^2+\ldots\ldots$$
$$+\{b_2t_1t_2+b_1(t_1-t_2)-b_0\}z^n$$
$$+\{b_1t_1t_2+b_0(t_1-t_2)\}z^{n+1}+b_0t_1t_2z^{n+2}\big].$$

Now

$$|\phi(z)|\le|a_n(t_1-t_2)-a_{n-1}||z|$$
$$+|a_nt_1t_2+a_{n-1}(t_1-t_2)-a_{n-2}||z|^2+\ldots\ldots$$

$$+|a_2t_1t_2+a_1(t_1-t_2)-a_0||z|^n$$
$$+|a_1t_1t_2+a_0(t_1-t_2)||z|^{n+1}$$
$$+|a_0t_1t_2||z|^{n+2}+|b_n(t_1-t_2)-b_{n-1}||z|$$
$$+|b_nt_1t_2+b_{n-1}(t_1-t_2)-b_{n-2}||z|^2+\ldots\ldots$$
$$+|b_2t_1t_2+b_1(t_1-t_2)-b_0||z|^n$$
$$+|b_1t_1t_2+b_0(t_1-t_2)||z|^{n+1}+|b_0t_1t_2||z|^{n+2}.$$

This gives after using hypothesis, for |z| = 1/t1

$$|\phi(z)| \le (a_n + b_n) + (|a_0| - a_0)\frac{t_2}{t_1^{n+1}} + (|b_0| - b_0)\frac{t_2}{t_1^{n+1}}.$$

Clearly $\phi(z)$.

Therefore, by Schwarz's Lemma

$$|\phi(z)| \le (a_n + b_n) + (|a_0| - a_0)\frac{t_2}{t_1^{n+1}} + (|b_0| - b_0)\frac{t_2}{t_1^{n+1}}|z|t_1$$

Hence from (3.1), if $M = \frac{t_2}{t_1^{n+1}}\left[(|a_0| - a_0) + (|b_0| - b_0)\right]$, then we get

$$|G(z)| \ge |\alpha_n| - |\phi(z)|$$
$$\ge |\alpha_n| - |\alpha_n + M||z|t_1 > 0,$$

If $|\alpha_n| \ge |\alpha_n + M||z|t_1$ for $|z| \le \frac{1}{t_1}$.

That is $|G(z)| > 0$, if $|z| \le \frac{|\alpha_n|}{|\alpha_n + M|t_1}$,

Thus in $|z| \le \frac{1}{t_1}$, $|G(z)| > 0$, if $|z| \le \frac{|\alpha_n|}{|\alpha_n + M|t_1}$.

Consequently, all the zeros of G(z) lie in $|z| \ge \frac{|\alpha_n|}{|\alpha_n + M|t_1}$.

As $F(z) = z^{n+2}G\left(\frac{1}{\bar{z}}\right)$, we conclude that all the zeros of F(z) and hence all the zeros of p(z) lie in

$$|z| \le \frac{|\alpha_n + M|}{|\alpha_n|}t_1.$$

This completes the proof of the theorem 1.5.

Proof of Theorem 1.6. Consider the polynomial

$$f(z)=(t_2+z)(t_1-z)p(z)$$
$$=\left\{t_1t_2+(t_1-t_2)z-z^2\right\}\left\{\begin{matrix}\alpha_nz^n+\alpha_{n-1}z^{n-1}\\+......+\alpha_1z+\alpha_0\end{matrix}\right\}$$

$$=-\alpha_nz^{n+2}+\left\{\alpha_n(t_1-t_2)-\alpha_{n-1}\right\}z^{n+1}$$
$$+\left\{\alpha_nt_1t_2+\alpha_{n-1}(t_1-t_2)-\alpha_{n-2}\right\}z^n+......$$
$$+\left\{\alpha_2t_1t_2+\alpha_1(t_1-t_2)-a_0\right\}z^2+$$
$$\left\{\alpha_1t_1t_2+\alpha_0(t_1-t_2)\right\}z+\alpha_0t_1t_2$$
$$=-\alpha_nz^{n+2}-\left\{\begin{matrix}k\alpha_n(t_1-t_2)-\alpha_n(t_1-t_2)\\-k\alpha_n(t_1-t_2)+\alpha_{n-1}\end{matrix}\right\}z^{n+1}$$
$$+\left\{\alpha_nt_1t_2+\alpha_{n-1}(t_1-t_2)-\alpha_{n-2}\right\}z^n+......$$
$$+\left\{\alpha_2t_1t_2+\alpha_1(t_1-t_2)-\alpha_0\right\}z^2$$
$$+\left\{\alpha_1t_1t_2+\alpha_0(t_1-t_2)\right\}z+\alpha_0t_1t_2$$
$$=-\alpha_nz^{n+2}-\left\{\begin{matrix}(k-1)\alpha_n(t_1-t_2)\\-k\alpha_n(t_1-t_2)-\alpha_{n-1}\end{matrix}\right\}z^{n+1}$$
$$+\left\{\alpha_nt_1t_2+\alpha_{n-1}(t_1-t_2)-\alpha_{n-2}\right\}z^n+......$$
$$+\left\{\alpha_2t_1t_2+\alpha_1(t_1-t_2)-\alpha_0\right\}z^2$$
$$+\left\{\alpha_1t_1t_2+\alpha_0(t_1-t_2)\right\}z+\alpha_0t_1t_2$$
$$=-a_nz^{n+2}-\left\{\begin{matrix}(k-1)a_n(t_1-t_2)\\-ka_n(t_1-t_2)+a_{n-1}\end{matrix}\right\}z^{n+1}$$
$$+\left\{a_nt_1t_2+a_{n-1}(t_1-t_2)-a_0\right\}z^2$$
$$+\left\{a_1t_1t_2+a_0(t_1-t_2)\right\}z+a_0t_1t_2$$
$$+i\left[-b_nz^{n+2}-\left\{\begin{matrix}(k-1)b_n(t_1-t_2)\\-kb_n(t_1-t_2)+b_{n-1}\end{matrix}\right\}z^{n+1}\right.$$
$$+\left\{b_nt_1t_2+b_{n-1}(t_1-t_2)-b_0\right\}z^2$$
$$\left.+\left\{b_1t_1t_2+b_0(t_1-t_2)\right\}z+b_0t_1t_2\right]$$

$$
\begin{aligned}
f(z) &= (t_2 + z)(t_1 - z)\,p(z) \\
&= \left\{ t_1 t_2 + (t_1 - t_2) z - z^2 \right\} \begin{Bmatrix} \alpha_n z^n + \alpha_{n-1} z^{n-1} \\ + \ldots\ldots + \alpha_1 z + \alpha_0 \end{Bmatrix} \\
&= -\alpha_n z^{n+2} + \left\{ \alpha_n (t_1 - t_2) - \alpha_{n-1} \right\} z^{n+1} \\
&\quad + \left\{ \alpha_n t_1 t_2 + \alpha_{n-1} (t_1 - t_2) - \alpha_{n-2} \right\} z^n + \ldots\ldots \\
&\quad + \left\{ \alpha_2 t_1 t_2 + \alpha_1 (t_1 - t_2) - a_0 \right\} z^2 + \\
&\quad \left\{ \alpha_1 t_1 t_2 + \alpha_0 (t_1 - t_2) \right\} z + \alpha_0 t_1 t_2 \\
&= -\alpha_n z^{n+2} - \begin{Bmatrix} k\alpha_n (t_1 - t_2) - \alpha_n (t_1 - t_2) \\ -k\alpha_n (t_1 - t_2) + \alpha_{n-1} \end{Bmatrix} z^{n+1} \\
&\quad + \left\{ \alpha_n t_1 t_2 + \alpha_{n-1} (t_1 - t_2) - \alpha_{n-2} \right\} z^n + \ldots\ldots \\
&\quad + \left\{ \alpha_2 t_1 t_2 + \alpha_1 (t_1 - t_2) - \alpha_0 \right\} z^2 \\
&\quad + \left\{ \alpha_1 t_1 t_2 + \alpha_0 (t_1 - t_2) \right\} z + \alpha_0 t_1 t_2 \\
&= -\alpha_n z^{n+2} - \begin{Bmatrix} (k-1)\alpha_n (t_1 - t_2) \\ -k\alpha_n (t_1 - t_2) - \alpha_{n-1} \end{Bmatrix} z^{n+1} \\
&\quad + \left\{ \alpha_n t_1 t_2 + \alpha_{n-1} (t_1 - t_2) - \alpha_{n-2} \right\} z^n + \ldots\ldots \\
&\quad + \left\{ \alpha_2 t_1 t_2 + \alpha_1 (t_1 - t_2) - \alpha_0 \right\} z^2 \\
&\quad + \left\{ \alpha_1 t_1 t_2 + \alpha_0 (t_1 - t_2) \right\} z + \alpha_0 t_1 t_2
\end{aligned}
$$

$$= -a_n z^{n+2} - \begin{Bmatrix} (k-1)a_n(t_1 - t_2) \\ -ka_n(t_1 - t_2) + a_{n-1} \end{Bmatrix} z^{n+1}$$

$$+\{a_n t_1 t_2 + a_{n-1}(t_1 - t_2) - a_0\} z^2$$

$$+\{a_1 t_1 t_2 + a_0(t_1 - t_2)\} z$$

$$+a_0 t_1 t_2 + i\left[-b_n z^{n+2} - \begin{Bmatrix} (k-1)b_n(t_1 - t_2) \\ -kb_n(t_1 - t_2) + b_{n-1} \end{Bmatrix} z^{n+1} \right.$$

$$+\{b_n t_1 t_2 + b_{n-1}(t_1 - t_2) - b_0\} z^2$$

$$+\{b_1 t_1 t_2 + b_0(t_1 - t_2)\} z + b_0 t_1 t_2]$$

$$= -a_n z^{n+2} - (k-1)a_n(t_1 - t_2) z^{n+1}$$

$$+\{ka_n(t_1 - t_2) - a_{n-1}\} z^{n+1}$$

$$+\{a_n t_1 t_2 + a_{n-1}(t_1 - t_2) - a_{n-2}\} z^n + \ldots\ldots$$

$$+\{a_2 t_1 t_2 + a_1(t_1 - t_2) + a_0\} z^2$$

$$+\{a_1 t_1 t_2 + a_0(t_1 - t_2)\} z + a_0 t_1 t_2$$

$$+i\left[-b_n z^{n+2} - (k-1)b_n(t_1 - t_2) z^{n+1} \right.$$

$$+\{b_n t_1 t_2 + b_{n-1}(t_1 - t_2) - b_{n-2}\} z^n + \ldots\ldots$$

$$+\{b_2 t_1 t_2 + b_1(t_1 - t_2) - b_0\} z^2$$

$$+\{b_1 t_1 t_2 + b_0(t_1 - t_2)\} z + b_0 t_1 t_2].$$

This gives

$$|f(z)| \geq |\alpha_n| |z|^{n+1} |z + (k-1)(t_1 - t_2)| - |\phi(z)|.$$

Therefore for $|z| > t_1$, we have

$$\left|f(z)\right| \geq |z|^{n+1}\Big[\left|\alpha_n\right|\left|z+(k-1)(t_1-t_2)\right|$$
$$-\left|ka_n(t_1-t_2)-a_{n-1}\right|$$
$$-\left|a_n t_1 t_2 + a_{n-1}(t_1-t_2)-a_{n-2}\right|\frac{1}{t_1}-\ldots\ldots$$
$$-\left|a_2 t_1 t_2 + a_1(t_1-t_2)-a_0\right|\frac{1}{t_1^{n-1}}$$
$$-\left|a_1 t_1 t_2 + a_0(t_1-t_2)\right|\frac{1}{t_1^{n}}-\left|a_0 t_1 t_2\right|\frac{1}{t_1^{n+1}}$$
$$-\left|kb_n(t_1-t_2)-b_{n-1}\right|$$
$$-\left|b_n t_1 t_2 + b_{n-1}(t_1-t_2)-b_{n-2}\right|\frac{1}{t_1}-\ldots\ldots$$
$$-\left|b_2 t_1 t_2 + b_1(t_1-t_2)-b_0\right|\frac{1}{t_1^{n-1}}$$
$$-\left|b_1 t_1 t_2 + b_0(t_1-t_2)\right|\frac{1}{t_1^{n}}-\left|b_0 t_1 t_2\right|\frac{1}{t_1^{n+1}}\Big]$$
$$=|z|^{n+1}\Big[\left|\alpha_n\right|\left|z+(k-1)(t_1-t_2)\right|$$
$$-\left\{\begin{matrix} ka_n(t_1-t_2)+a_n t_2 \\ -\dfrac{a_1 t_2}{t_1^{n-1}}-\dfrac{a_0}{t_1^{n-1}} \end{matrix}\right\}-\left\{\begin{matrix} kb_n(t_1-t_2)+b_n t_2 \\ -\dfrac{b_1 t_2}{t_1^{n-1}}-\dfrac{b_0}{t_1^{n-1}} \end{matrix}\right\}$$

$$-\left|a_1 t_1 t_2 + a_0(t_1-t_2)\right|\frac{1}{t_1^{n}}$$
$$-\left|b_1 t_1 t_2 + b_0(t_1-t_2)\right|\frac{1}{t_1^{n}}-\left|a_0 t_1 t_2\right|\frac{1}{t_1^{n+1}}-\left|b_0 t_1 t_2\right|\frac{1}{t_1^{n+1}}\Big]$$
$$>0,$$

if

$$
\begin{aligned}
&\left|z+(k-1)(t_1-t_2)\right| \\
&> \frac{1}{|\alpha_n|}\Bigg[\left\{ka_n(t_1-t_2)+a_nt_2-\frac{a_1t_2}{t_1^{n-1}}-\frac{a_0}{t_1^{n-1}}\right\} \\
&\quad +\left\{kb_n(t_1-t_2)+b_nt_2-\frac{b_1t_2}{t_1^{n-1}}-\frac{b_0}{t_1^{n-1}}\right\} \\
&\quad +\left|a_1t_1t_2+a_0(t_1-t_2)\right|\frac{1}{t_1^n} \\
&\quad +\left|b_1t_1t_2+b_0(t_1-t_2)\right|\frac{1}{t_1^n}+\left(|a_0|+|b_0|\right)\frac{t_2}{t_1^n}\Bigg] \\
&= R_1.
\end{aligned}
$$

Hence all the zeros of $f(z)$ whose modulus is greater than t_1 lie in the circle

$$\left|z+(k-1)(t_1-t_2)\right| \le R_1.$$

Since all the zeros whose modulus is less than t_1 already lies in this circle, we conclude that all the zeros of $f(z)$ and hence $p(z)$ lie in

$$\left|z+(k-1)(t_1-t_2)\right| \le R_1.$$

This completes the proof of the Theorem 1.6.

REFERENCES

1. A. Aziz and Q. G. Mohammad, Zero free regions for polynomials and some generalizations of Enström-Kakeya theorem, Cand. Math. Bull., 27(1984), 265-272.
2. G. Eneström, Remarqueesur un The'ore'me'relatif aux racines de I' equation anxn + an-1xn-1 + ...+ a0 = 0outous les coefficients sont reels et possitifs, Tôhoku Math. J., 18 (1920), 34-36.
3. A. Hurwitz, Űberdie Nullslenllen der Bessel Schen Function, Math. Ann., 33(1989)246-266.

4. A.Joyal, G.Labelle and Q.I. Rahman, On the location of zeros of polynomials, Canad. Math. Bull., 10 (1967), 53-63.

5. S.Kakeya, On the limits of the roots of an algebraic equation with positive coefficients, Tôhoku Math. J., 2 (1912-13), 140-142.

6. A. Aziz and B. A. Zargar, Some extensions of Enström-Kakeya Theorem, Glasnik Mate., 31(1996), 239-244.

16

MODULAR RELATIONS FOR THE SEXTODECIC ANALOGUES OF THE ROGERS-RAMANUJAN FUNCTIONS WITH ITS APPLICATIONS TO PARTITIONS

Adiga Chandrashekar[1], Nasser Abdo Saeed Bulkhali

[1] Department of Studies in Mathematics, University of Mysore, Manasagangotri, Mysore, IndiaKlong Luang, Pathumthani 12120, Thailand

ABSTRACT

In his Ph.D. thesis, C. Gugg considered four functions of order 16 that are analogues of the Rogers-Ramanujan functions and established 12 modular relations involving these functions. In this paper, we obtain 16 new modular relations for these functions. Furthermore, we give partition theoretic interpretations for some of our modular relations.

INTRODUCTION

Throughout the paper, we assume $|q|<1$. We use the standard notation

$$(a;q)_0 := 1, \qquad (a;q)_n := \prod_{j=0}^{n-1}(1-aq^j)$$

And $(a;q)_\infty := \prod_{n=0}^{\infty}(1-aq^n).$

Ramanujan's general theta function is defined by

$$f(a,b) := \sum_{n=-\infty}^{\infty} a^{n(n+1)/2} b^{n(n-1)/2}, \qquad |ab| < 1. \tag{1}$$

The Jacobi triple product identity [[1], Entry 19] in Ramanujan's notation is

$$f(a,b) = (-a;ab)_\infty (-b;ab)_\infty (ab;ab)_\infty. \tag{2}$$

Ramanujan defined the following three special cases of (1) [[1], Entry 22]:

$$\phi(q) := f(q,q) = \sum_{n=-\infty}^{\infty} q^{n^2} = (-q;q^2)_\infty^2 (q^2;q^2)_\infty, \tag{3}$$

$$\psi(q) := f(q,q^3) = \sum_{n=0}^{\infty} q^{n(n+1)/2} = \frac{(q^2;q^2)_\infty}{(q;q^2)_\infty}, \tag{4}$$

and

$$f(-q) := f(-q,-q^2) = \sum_{n=-\infty}^{\infty} (-1)^n q^{n(3n-1)/2} = (q;q)_\infty \tag{5}$$

For convenience, we define

$$f_n := f(-q^n) = (q^n;q^n)_\infty,$$

for a positive integer n.

The well-known Rogers-Ramanujan functions are defined by

$$G(q) := \sum_{n=0}^{\infty} \frac{q^{n^2}}{(q;q)_n}$$

And

$$H(q) := \sum_{n=0}^{\infty} \frac{q^{n(n+1)}}{(q;q)_n} \tag{6}$$

These functions satisfy the famous Rogers-Ramanujan identities

$$G(q) = \frac{1}{(q;q^5)_\infty (q^4;q^5)_\infty}$$

and

$$H(q) = \frac{1}{(q^2;q^5)_\infty (q^3;q^5)_\infty}. \tag{7}$$

In[27], Ramanujan remarks, "I have now found an algebraic relation between $G(q)$ and $H(q)$, viz.:

$$H(q)\{G(q)\}^{11} - q^2 G(q)\{H(q)\}^{11} = 1 + 11q\{G(q)H(q)\}^6.$$

Another noteworthy formula is

$$H(q)G(q^{11}) - q^2 G(q)H(q^{11}) = 1.$$

Each of these formulae is the simplest of a large class." In his lost notebook[28], Ramanujan recorded forty beautiful modular relations involving the Rogers-Ramanujan functions without proofs. The forty identities were first brought before the mathematical world by B. J. Birch[18]. Many of these identities have been established by L. J. Rogers[30], G. N. Watson [34], D. Bressoud [20,21] A. J. F. Biagioli [17] and B. C. Berndt et al. [16] offered proofs of 35 of the 40 identities. Recently, in Chapter 8 of their book[11], G. E. Andrews and Berndt collected proofs for all forty identities. Most likely these proofs might have given by Ramanujan himself. A number of mathematicians tried to find new identities for the Rogers-Ramanujan functions similar to those which have been found by Ramanujan[28], including Berndt and H. Yesilyurt [15], Yesilyurt [36], S. Robins [29] and C. Gugg[23].

Two beautiful analogues of the Rogers-Ramanujan functions are the Göllnitz-Gordon functions, which are defined as

$$S(q) := \sum_{n=0}^{\infty} \frac{(-q;q^2)_n}{(q^2;q^2)_n} q^{n^2} = \frac{1}{(q,q^4,q^7;q^8)_\infty} \tag{8}$$

And

$$T(q) := \sum_{n=0}^{\infty} \frac{(-q;q^2)_n}{(q^2;q^2)_n} q^{n^2+2n} = \frac{1}{(q^3,q^4,q^5;q^8)_\infty}. \tag{9}$$

Identities (8) and (9) can be found in L. J. Slater's list [31]. S.-S. Huang [26] has established a number of modular relations for $S(q)$ and $T(q)$. S.-L. Chen and Huang[22] have derived some new modular relations involving $S(q)$ and $T(q)$. N. D. Baruah, J. Bora and N. Saikia[14] offered new proofs of many of these modular relations, as well as establishing some new relations. E. X. W. Xia and X. M. Yao[35] offered new proofs of some modular relations established by Huang [26] and Chen and Huang[22]. They also established some new relations that involve only Göllnitz-Gordon functions.

In view of the Ramanujan's forty identities, many of the Rogers-Ramanujan type functions were studied by many mathematicians. For example, septic analogues of the Rogers-Ramanujan functions were studied by H. Hahn [24, 25], nonic analogues of the Rogers-Ramanujan functions were studied by Baruah and Bora[13], cubic functions were studied by C. Adiga et al.[4,5,9], another cubic functions were studied by K. R. Vasuki, G. Sharath and K. R. Rajanna [33], dodecic analogues of the Rogers-Ramanujan functions were studied by Baruah and Bora [12], Robins [29] and C. Gugg [23], another dodecic analogues of the Rogers-Ramanujan functions were studied by Vasuki and P. S. Guruprasad [32] Adiga, Vasuki and B. R. Srivatsa Kumar [10] established modular relations involving two functions of Rogers-Ramanujan type, the authors have studied two functions of order ten [2,3] and more recently, Adiga et al. [6, 7, 8] have studied four functions of order fifteen. Almost all of these functions which have been studied so far can be found in Slater's list [31].

In Chapter 7 of his Ph.D. thesis, C. Gugg [23] considered the following four sextodecic analogues of the Rogers-Ramanujan functions:

$$I(q) := \frac{(q^7,q^9,q^{16};q^{16})_\infty}{(q;q)_\infty} = \frac{f(-q^7,-q^9)}{f(-q)}, \tag{10}$$

$$J(q) := \frac{(q^5,q^{11},q^{16};q^{16})_\infty}{(q;q)_\infty} = \frac{f(-q^5,-q^{11})}{f(-q)}, \tag{11}$$

$$K(q) := \frac{(q^3,q^{13},q^{16};q^{16})_\infty}{(q;q)_\infty} = \frac{f(-q^3,-q^{13})}{f(-q)} \tag{12}$$

and

Gugg [23] established twelve modular relations involving the functions $I(q)$, $J(q)$, $K(q)$ and $L(q)$ including the following two beautiful identities:

$$I_1I_3 + qJ_1J_3 + q^3K_1K_3 + q^6L_1L_3 = \frac{f_2^2 f_6^2}{f_1^2 f_3 f_{12}}, \tag{14}$$

$$I_1I_7 + q^2J_1J_7 + q^6K_1K_7 + q^{12}L_1L_7 = \frac{f_4^2 f_7}{f_1 f_2 f_{14}}. \tag{15}$$

To prove his results, Gugg [23] applied a theorem of R. Blecksmith, J. Brillhart and I. Gerst [19] and also employed the method given by Bressoud in his thesis [20].

The main purpose of this paper is to establish several modular relations that are analogues of Ramanujan's forty identities involving $I(q)$, $J(q)$, $K(q)$ and $L(q)$. Many of these identities that we find have partition theoretic interpretations. In the last section, we extract partition theoretic interpretations for some of these modular relations. To prove our results, we use the idea of Watson [34] which he has used to prove some of Ramanujan's forty identities.

MAIN RESULTS

In this section, we present a list of new modular relations for the functions $I(q)$, $J(q)$, $K(q)$ and $L(q)$, which we establish in Section 4. For simplicity, we use the notations $I_n := I(q^n)$, $J_n := J(q^n)$, $K_n := K(q^n)$ and $L_n := L(q^n)$, for a positive integer n.

$$I_2^2 + qJ_2^2 + q^3K_2^2 + q^6L_2^2 = \frac{f_8^2}{f_1 f_{16}}, \tag{16}$$

$$I_3I_5 + q^2J_3J_5 + q^6K_3K_5 + q^{12}L_3L_5 = \frac{f_4^2 f_{15}^2}{f_2 f_3 f_5 f_{30}}, \tag{17}$$

$$I_3L_1 - J_3K_1 + qK_3J_1 - q^3L_3I_1 = 0, \tag{18}$$

$$I_3J_1 + q^2J_3L_1 + q^2K_3I_1 - q^5L_3K_1 = \frac{f_4^5 f_{12}}{f_1 f_2^{\,2} f_6 f_8^{\,2}}, \tag{19}$$

$$I_7L_1 - qJ_7K_1 + q^4K_7J_1 - q^9L_7I_1 = \frac{f_1 f_{28}^2}{f_2 f_7 f_{14}}, \tag{20}$$

$$-qI_5L_3 + J_5K_3 - qK_5J_3 + q^4L_5I_3 = \frac{f_1^2 f_{60}^2}{f_2 f_3 f_5 f_{30}}, \tag{21}$$

$$I_3L_1 + J_3K_1 - qK_3J_1 - q^3L_3I_1 = 2\frac{f_4 f_{24}^2}{f_2 f_3 f_{12}}, \tag{22}$$

$$I_3K_1 + J_3I_1 - q^3K_3L_1 + q^4L_3J_1 = 2\frac{f_8 f_{12}}{f_1 f_4 f_6}, \tag{23}$$

$$\begin{aligned} &I_{15}I_1 + q^4J_{15}J_1 + q^{12}K_{15}K_1 + q^{24}L_{15}L_1 \\ &\quad = \frac{1}{f_1 f_{15}}\left\{\frac{f_8^2 f_{60}^5}{f_4 f_{30}^2 f_{120}^2} - q^7\frac{f_4^5 f_{120}^2}{f_2^2 f_8^2 f_{60}}\right\}, \end{aligned} \tag{24}$$

$$\begin{aligned} &I_{13}I_3 + q^4J_{13}J_3 + q^{12}K_{13}K_3 + q^{24}L_{13}L_3 \\ &\quad = \frac{1}{f_3 f_{13}}\left\{\frac{f_8^2 f_{156}^5}{f_4 f_{78}^2 f_{312}^2} - q^{19}\frac{f_4^5 f_{312}^2}{f_2^2 f_8^2 f_{156}}\right\}, \end{aligned} \tag{25}$$

$$\begin{aligned} &I_{11}I_5 + q^4J_{11}J_5 + q^{12}K_{11}K_5 + q^{24}L_{11}L_5 \\ &\quad = \frac{1}{f_5 f_{11}}\left\{\frac{f_8^2 f_{220}^5}{f_4 f_{110}^2 f_{440}^2} - q^{27}\frac{f_4^5 f_{440}^2}{f_2^2 f_8^2 f_{220}}\right\}, \end{aligned} \tag{26}$$

$$I_9I_7+q^4J_9J_7+q^{12}K_9K_7+q^{24}L_9L_7$$
$$=\frac{1}{f_7f_9}\left\{\frac{f_8^2f_{252}^5}{f_4f_{126}^2f_{504}^2}-q^{31}\frac{f_4^5f_{504}^2}{f_2^2f_8^2f_{252}}\right\},\tag{27}$$

$$I_{14}J_2+q^6J_{14}L_2+q^{10}K_{14}I_2-q^{22}L_{14}K_2$$
$$=\frac{1}{2f_2f_{14}}\left\{\frac{f_2^2f_7f_{28}}{f_1f_{14}}+\frac{f_1f_4f_{14}^2}{f_2f_7}\right\},\tag{28}$$

$$I_{14}K_2+q^2J_{14}I_2-q^{12}K_{14}L_2+q^{20}L_{14}J_2$$
$$=\frac{1}{2qf_2f_{14}}\left\{\frac{f_2^2f_7f_{28}}{f_1f_{14}}-\frac{f_1f_4f_{14}^2}{f_2f_7}\right\},\tag{29}$$

$$I_{10}J_6+q^{10}J_{10}L_6+q^6K_{10}I_6-q^{18}L_{10}K_6$$
$$=\frac{1}{2f_6f_{10}}\left\{\frac{f_2^2f_{15}f_{60}}{f_1f_{30}}+\frac{f_1f_4f_{30}^2}{f_2f_{15}}\right\},\tag{30}$$

$$I_6J_{10}+q^{14}J_6L_{10}+q^2K_6I_{10}-q^{14}L_6K_{10}$$
$$=\frac{1}{2qf_6f_{10}}\left\{\frac{f_2^2f_{15}f_{60}}{f_1f_{30}}-\frac{f_1f_4f_{30}^2}{f_2f_{15}}\right\}.\tag{31}$$

Identity (18) is the corrected version of identity (7.2.12) found in [23].

The following two identities are relations involving some combinations of the functions defined in (10)–(13), and the Göllnitz-Gordon functions $S(q)$ and $T(q)$:

$$\frac{I_1J_1-q^2K_1L_1}{S(-q)}=\frac{f_2^3f_8}{f_1^3f_4},\tag{32}$$

$$\frac{I_1K_1+qJ_1L_1}{T(-q)}=\frac{f_2^{\,3}f_8}{f_1^{\,3}f_4}. \tag{33}$$

SOME PRELIMINARY RESULTS

The function $f(a,b)$ satisfies the following basic properties [1]:

$$f(a,b)=f(b,a), \tag{34}$$

$$f(1,a)=2f(a,a^3), \tag{35}$$

$$f(-1,a)=0, \tag{36}$$

and, if n is an integer,

$$f(a,b)=a^{n(n+1)/2}b^{n(n-1)/2}f(a(ab)^n,b(ab)^{-n}). \tag{37}$$

Lemma 3.1 We have

$$\phi(q)=\frac{f_2^5}{f_1^2f_4^2},\qquad \psi(q)=\frac{f_2^2}{f_1},$$

$$\phi(-q)=\frac{f_1^2}{f_2}\text{ and }\psi(-q)=\frac{f_1f_4}{f_2}.$$

This lemma is a consequence of the Jacobi triple product identity (2) and Entry 24 of [1].

The following identity follows easily from Entry 31 found in [1]

$$\phi(q^4)+2q\psi(q^8)=\phi(q). \tag{38}$$

Using (38) one can easily establish the following lemma:

Lemma 3.2 We have

$$\begin{aligned}&\phi(-q^a)\phi(q^b)-\phi(q^a)\phi(-q^b)\\&=4q^b\left\{\phi(q^{4a})\psi(q^{8b})-q^{a-b}\phi(q^{4b})\psi(q^{8a})\right\}.\end{aligned} \tag{39}$$

The following lemma can be found in [1, Entry 30(ii) and (iii)]:

Lemma 3.3 We have

$$f(a,b)+f(-a,-b)=2f(a^3b,ab^3), \tag{40}$$

$$f(a,b)-f(-a,-b)=2af(b/a,a^5b^3). \tag{41}$$

Lemma 3.4 Let $m=[s/(s-r)]$ $l=m(s-r)-r$, $k=-m(s-r)+s$ and $h=mr-m(m-1)(s-r)/2$, $0\leq r<s$. Here $[x]$ denote the largest integer less than or equal to x . Then,

$$f(-q^{-r},-q^s)=(-1)^m q^{-h} f(-q^l,-q^k)$$

For a proof of Lemma, 3.4 see [3].

PROOFS OF THE MAIN RESULTS

We prove our main results using ideas similar to those of Watson [34]. In all proofs, one expresses the left sides of the identities in terms of theta functions by using (10), (11), (12) and (13). After clearing fractions, we see that the right side can be expressed as a product of two theta functions, say with summations indices m and n. One then tries to find a change of indices of the form

$$\alpha m+\beta n=4M+a \quad \text{and} \quad \gamma m+\delta n=4N+b,$$

$$\alpha m+\beta n=8M+a \quad \text{and} \quad \gamma m+\delta n=8N+b$$

Or

$$\alpha m+\beta n=16M+a \quad \text{and} \quad \gamma m+\delta n=16N+b,$$

so that the product on the right side decomposes into the requisite sum of two products of theta functions on the left side.

Proof of (14). Using (10), (11), (12), (13) and Lemma 3.1, we see that (14) is equivalent to

$$\begin{aligned}\phi\left(-q^{6}\right)\psi(q)=qf\left(-q^{15},-q^{33}\right)f\left(-q^{5},-q^{11}\right)\\+f\left(-q^{27},-q^{21}\right)f\left(-q^{9},-q^{7}\right)\\+q^{3}f\left(-q^{39},-q^{9}\right)f\left(-q^{13},-q^{3}\right)\\+q^{6}f\left(-q^{45},-q^{3}\right)f\left(-q^{15},-q\right).\end{aligned} \tag{42}$$

We have

$$\begin{aligned}\phi\left(-q^{6}\right)\psi(q)&=f(-q^{6},-q^{6})f(q,q^{3})\\&=\sum_{m,n=-\infty}^{\infty}(-1)^{m}q^{6m^{2}+2n^{2}+n}.\end{aligned} \tag{43}$$

In this representation, we make the change of indices by setting

$m+n=4M+a$ and $-3m+n=4N+b,$

where a and b have values selected from the set $\{0,\pm 1,2\}$. Then

$m=M-N+(a-b)/4$ and $n=3M+N+(3a+b)/4$

It follows easily that $a=b$, and so $m=M-N$ and $n=3M+N+a$, where $-1\le a\le 2$ Thus, there is one-to-one correspondence between the set of all pairs of integers (m,n), $-\infty<m,n<\infty$, and triples of integers (M,N,a), $-\infty<M,N<\infty$ $-1\le a\le 2$ From (43), we find that

$$\phi\left(-q^6\right)\psi(q)=\sum_{a=-1}^{2} q^{2a^2+a}$$
$$\times\sum_{M=-\infty}^{\infty}(-1)^M q^{24M^2+(3+12a)M}$$
$$\times\sum_{N=-\infty}^{\infty}(-1)^N q^{8N^2+(1+4a)N}$$
$$=\sum_{a=-1}^{2} q^{2a^2+a} f\left(-q^{27+12a},-q^{21-12a}\right)$$
$$\times f\left(-q^{9+4a},-q^{7-4a}\right)$$
$$=qf\left(-q^{15},-q^{33}\right)f\left(-q^{5},-q^{11}\right)$$
$$+f\left(-q^{27},-q^{21}\right)f\left(-q^{9},-q^{7}\right)$$
$$+q^3 f\left(-q^{39},-q^{9}\right)f\left(-q^{13},-q^{3}\right)$$
$$+q^{10} f\left(-q^{51},-q^{-3}\right)f\left(-q^{17},-q^{-1}\right),$$

which is same as (42).

Proof of(15). Using (10), (11), (12), (13) and Lemma 3.1, we see that (15) is equivalent to

$$\begin{aligned}\phi\left(-q^7\right)\psi\left(q^2\right)&=f\left(-q^{63},-q^{49}\right)f\left(-q^{9},-q^{7}\right)\\&+q^2 f\left(-q^{77},-q^{35}\right)f\left(-q^{11},-q^{5}\right)\\&+q^6 f\left(-q^{91},-q^{21}\right)f\left(-q^{13},-q^{3}\right)\\&+q^{12} f\left(-q^{105},-q^{7}\right)f\left(-q^{15},-q\right).\end{aligned} \tag{44}$$

We have

$$\begin{aligned}2\phi\left(-q^7\right)\psi\left(q^2\right)&=f(-q^7,-q^7)f(1,q^2)\\&=\sum_{m,n=-\infty}^{\infty}(-1)^m q^{7m^2+n^2+n}.\end{aligned} \tag{45}$$

In these representations, we make the change of indices by setting

$$m+n=8M+a \text{ and } -7m+n=8N+b.$$

where a and b have values selected from the set $\{0,\pm1,\pm2,\pm3,4\}$. Then

$$m=M-N+(a-b)/8 \text{ and } n=7M+N+(7a+b)/8.$$

It follows easily that $a=b$, and so $m=M-N$ and $n=7M+N+a$, where $-3\le a\le 4$ Thus, there is one-to-one correspondence between the set of all pairs of integers (m,n), $-\infty<m,n<\infty$, and triples of integers (M,N,a), $-\infty<M,N<\infty$, $-3\le a\le 4$. From (45), we find that

$$\begin{aligned}2\phi\left(-q^7\right)\psi\left(q^2\right) &= \sum_{a=-3}^{4} q^{a^2+a} \\ &\quad\times \sum_{M=-\infty}^{\infty} (-1)^M q^{56M^2+(7+14a)M} \\ &\quad\times \sum_{N=-\infty}^{\infty} (-1)^N q^{16N^2+(1+2a)N} \\ &= \sum_{a=-3}^{4} q^{a^2+a} f\left(-q^{63+14a},-q^{49-14a}\right) \\ &\quad\times f\left(-q^{9+2a},-q^{7-2a}\right) \\ &= q^6 f\left(-q^{21},-q^{91}\right) f\left(-q^3,-q^{13}\right) \\ &\quad+q^2 f\left(-q^{35},-q^{77}\right) f\left(-q^5,-q^{11}\right) \\ &\quad+f\left(-q^{49},-q^{63}\right) f\left(-q^7,-q^9\right) \\ &\quad+f\left(-q^{63},-q^{49}\right) f\left(-q^9,-q^7\right) \\ &\quad+q^2 f\left(-q^{77},-q^{35}\right) f\left(-q^{11},-q^5\right) \\ &\quad+q^6 f\left(-q^{91},-q^{21}\right) f\left(-q^{13},-q^3\right) \\ &\quad+q^{12} f\left(-q^{105},-q^7\right) f\left(-q^{15},-q\right) \\ &\quad+q^{20} f\left(-q^{119},-q^{-7}\right) f\left(-q^{17},-q^{-1}\right),\end{aligned}$$

Now changing q to q^2, and then applying Lemma 3.2 in the resulting identity, we may rewrite (46) in the form

$$\begin{aligned}\phi\left(q^{30}\right)\psi\left(q^{4}\right)-q^{7}\phi\left(q^{2}\right)\psi\left(q^{60}\right) &= f\left(-q^{135},-q^{105}\right)f\left(-q^{9},-q^{7}\right)\\ &+q^{4}f\left(-q^{165},-q^{75}\right)f\left(-q^{11},-q^{5}\right)\\ &+q^{12}f\left(-q^{195},-q^{45}\right)f\left(-q^{13},-q^{3}\right)\\ &+q^{24}f\left(-q^{225},-q^{15}\right)f\left(-q^{15},-q\right).\end{aligned} \tag{46}$$

$$\begin{aligned}\frac{1}{4q}\left\{\phi\left(-q^{15}\right)\phi(q)-\phi\left(q^{15}\right)\phi(-q)\right\} &= f\left(-q^{270},-q^{210}\right)f\left(-q^{18},-q^{14}\right)\\ &+q^{8}f\left(-q^{330},-q^{150}\right)f\left(-q^{22},-q^{10}\right)\\ &+q^{24}f\left(-q^{390},-q^{90}\right)f\left(-q^{26},-q^{6}\right)\\ &+q^{48}f\left(-q^{450},-q^{30}\right)f\left(-q^{30},-q^{2}\right).\end{aligned} \tag{47}$$

Thus we need only to establish (47). We have

$$\begin{aligned}\phi\left(-q^{15}\right)\phi(q) &= f(-q^{15},-q^{15})f(q,q)\\ &= \sum_{m,n=-\infty}^{\infty}(-1)^{m}q^{15m^2+n^2}.\end{aligned} \tag{48}$$

In these representations, we make the change of indices by setting

$$m+n=16M+a \quad \text{and} \quad -15m+n=16N+b,$$

where a and b have values selected from the set $\{0,\pm1,\pm2,\pm3,\pm4,\pm5,\pm6,\pm7,8\}$. Then

$$m=M-N+(a-b)/16 \text{ and } n=15M+N+(15a+b)/16$$

It follows easily that $a=b$, and so $m=M-N$ and $n=15M+N+a$, where $-7\leq a\leq 8$. Thus, there is one-to-one correspondence between the set of all pairs of integers $(m,n), -\infty<m,n<\infty$, and triples of integers $(M,N,a), -\infty<M,N<\infty$, $-7\leq a\leq$ From (48), we find that

$$\phi\left(-q^{15}\right)\phi(q)=\sum_{a=-7}^{8}q^{a^2}\sum_{M=-\infty}^{\infty}(-1)^M q^{240M^2+30aM}$$

$$\times\sum_{N=-\infty}^{\infty}(-1)^N q^{16N^2+2aN}$$

$$=\sum_{a=-7}^{8}q^{a^2}f\left(-q^{240+30a},-q^{240-30a}\right)$$

$$\times f\left(-q^{16+2a},-q^{16-2a}\right)$$

$$=q^{49}f\left(-q^{30},-q^{450}\right)f\left(-q^{2},-q^{30}\right)$$
$$+q^{36}f\left(-q^{60},-q^{420}\right)f\left(-q^{4},-q^{28}\right)$$
$$+q^{25}f\left(-q^{90},-q^{390}\right)f\left(-q^{6},-q^{26}\right)$$
$$+q^{16}f\left(-q^{120},-q^{360}\right)f\left(-q^{8},-q^{24}\right)$$
$$+q^{9}f\left(-q^{150},-q^{330}\right)f\left(-q^{10},-q^{22}\right)$$
$$+q^{4}f\left(-q^{180},-q^{300}\right)f\left(-q^{12},-q^{20}\right)$$
$$+qf\left(-q^{210},-q^{270}\right)f\left(-q^{14},-q^{18}\right)$$
$$+f\left(-q^{240},-q^{240}\right)f\left(-q^{16},-q^{16}\right) \tag{49}$$
$$+qf\left(-q^{270},-q^{210}\right)f\left(-q^{18},-q^{14}\right)$$
$$+q^{4}f\left(-q^{300},-q^{180}\right)f\left(-q^{20},-q^{12}\right)$$
$$+q^{9}f\left(-q^{330},-q^{150}\right)f\left(-q^{22},-q^{10}\right)$$
$$+q^{16}f\left(-q^{360},-q^{120}\right)f\left(-q^{24},-q^{8}\right)$$
$$+q^{25}f\left(-q^{390},-q^{90}\right)f\left(-q^{26},-q^{6}\right)$$
$$+q^{36}f\left(-q^{420},-q^{60}\right)f\left(-q^{28},-q^{4}\right)$$
$$+q^{49}f\left(-q^{450},-q^{30}\right)f\left(-q^{30},-q^{2}\right)$$
$$+q^{64}f\left(-q^{480},-1\right)f\left(-q^{32},-1\right).$$

Using the same change of indices for the product $\phi(q^{15})\phi(-q)$, we find that

$$\phi(q^{15})\phi(-q)=\sum_{a=-7}^{8}(-1)^a q^{a^2}$$
$$\times\sum_{M=-\infty}^{\infty}(-1)^M q^{240M^2+30aM}$$
$$\times\sum_{N=-\infty}^{\infty}(-1)^N q^{16N^2+2aN}$$
$$=\sum_{a=-7}^{8}(-1)^a q^{a^2} f\left(-q^{240+30a},-q^{240-30a}\right)$$
$$\times f\left(-q^{16+2a},-q^{16-2a}\right)$$
$$=-q^{49} f\left(-q^{30},-q^{450}\right) f\left(-q^{2},-q^{30}\right)$$
$$+q^{36} f\left(-q^{60},-q^{420}\right) f\left(-q^{4},-q^{28}\right)$$
$$-q^{25} f\left(-q^{90},-q^{390}\right) f\left(-q^{6},-q^{26}\right)$$
$$+q^{16} f\left(-q^{120},-q^{360}\right) f\left(-q^{8},-q^{24}\right)$$
$$-q^{9} f\left(-q^{150},-q^{330}\right) f\left(-q^{10},-q^{22}\right)$$
$$+q^{4} f\left(-q^{180},-q^{300}\right) f\left(-q^{12},-q^{20}\right)$$

$$\begin{aligned}
&-qf\left(-q^{210},-q^{270}\right)f\left(-q^{14},-q^{18}\right)\\
&+f\left(-q^{240},-q^{240}\right)f\left(-q^{16},-q^{16}\right)\\
&-qf\left(-q^{270},-q^{210}\right)f\left(-q^{18},-q^{14}\right)\\
&+q^{4}f\left(-q^{300},-q^{180}\right)f\left(-q^{20},-q^{12}\right)\\
&-q^{9}f\left(-q^{330},-q^{150}\right)f\left(-q^{22},-q^{10}\right)\\
&+q^{16}f\left(-q^{360},-q^{120}\right)f\left(-q^{24},-q^{8}\right)\\
&-q^{25}f\left(-q^{390},-q^{90}\right)f\left(-q^{26},-q^{6}\right)\\
&+q^{36}f\left(-q^{420},-q^{60}\right)f\left(-q^{28},-q^{4}\right)\\
&-q^{49}f\left(-q^{450},-q^{30}\right)f\left(-q^{30},-q^{2}\right)\\
&+q^{64}f\left(-q^{480},-1\right)f\left(-q^{32},-1\right).
\end{aligned} \tag{50}$$

Subtracting (50) from (49), we deduce the desired result.

In a similar way, one can prove the identities (25)–(31).

Proof of(32) and (33). Using (10), (11), (12), (13), Lemma 3.1 and $S(-q)=\dfrac{f(q^3,q^5)}{\psi(q)}, T(-q)=\dfrac{f(q,q^7)}{\psi(q)}$, we see that (32) and (33) are equivalent, respectively, to

$$\begin{aligned}
\psi\left(-q^{2}\right)f\left(q^{3},q^{5}\right)&=f\left(-q^{7},-q^{9}\right)f\left(-q^{11},-q^{5}\right)\\
&\quad-q^{2}f\left(-q^{15},-q\right)f\left(-q^{13},-q^{3}\right)
\end{aligned} \tag{51}$$

and

$$\begin{aligned}
\psi\left(-q^{2}\right)f\left(q,q^{7}\right)&=f\left(-q^{3},-q^{13}\right)f\left(-q^{7},-q^{9}\right)\\
&\quad+qf\left(-q^{11},-q^{5}\right)f\left(-q^{15},-q\right).
\end{aligned} \tag{52}$$

Applying Lemma 3.3, we obtain

$$2f\left(q^6,q^{10}\right)=\psi(q)+\psi(-q)$$

and

$$2qf\left(q^2,q^{14}\right)=\psi(q)-\psi(-q)$$

We may rewrite (51) and (52) in the form

$$\begin{aligned}\psi\left(-q^4\right)\{\psi(q)+\psi(-q)\}\\ =2f\left(-q^{14},-q^{18}\right)f\left(-q^{22},-q^{10}\right)\\ -2q^4f\left(-q^{30},-q^2\right)f\left(-q^{26},-q^6\right),\end{aligned} \tag{53}$$

and

$$\begin{aligned}\psi\left(-q^4\right)\{\psi(q)-\psi(-q)\}\\ =2qf\left(-q^6,-q^{26}\right)f\left(-q^{14},-q^{18}\right)\\ +2q^3f\left(-q^{22},-q^{10}\right)f\left(-q^{30},-q^2\right).\end{aligned} \tag{54}$$

Thus to establish (32) and (33), it is suffices to prove (53) and (54).
We have

$$\begin{aligned}\psi(-q^4)\psi(q)&=f(-q^4,-q^{12})f(q,q^3)\\ &=\sum_{m,n=-\infty}^{\infty}(-1)^m q^{8m^2-4m+2n^2+n}.\end{aligned} \tag{55}$$

In this representation, we make the change of indices by setting

$$2m+n=4M+a \text{ and } -2m+n=4N+b,$$

where a and b have values selected from the set $\{0,\pm1,2\}$.
Then

$$m = M - N + (a-b)/4 \text{ and } n = 2M + N + (a+b)/2$$

It follows easily that $a=b$, and so $m=M-N$ and $n=2M+N+a$, where $-1 \le a \le 2$. Thus, there is one-to-one correspondence between the set of all pairs of integers (m,n), $-\infty < m,n < \infty$, and triples of integers (M,N,a), $-\infty < M,N < \infty, -1 \le a \le 2$. From (55), we find that

$$\begin{aligned}
\psi(-q^4)\psi(q) &= \sum_{a=-1}^{2} q^{2a^2+a} \\
&\quad \times \sum_{M=-\infty}^{\infty} (-1)^M q^{16M^2+(-2+8a)M} \\
&\quad \times \sum_{N=-\infty}^{\infty} (-1)^N q^{16N^2+(6+8a)N} \\
&= \sum_{a=-1}^{2} q^{2a^2+a} f\left(-q^{14+8a}, -q^{18-8a}\right) \\
&\quad \times f\left(-q^{22+8a}, -q^{10-8a}\right) \\
&= qf\left(-q^6, -q^{26}\right) f\left(-q^{14}, -q^{18}\right) \\
&\quad + f\left(-q^{14}, -q^{18}\right) f\left(-q^{22}, -q^{10}\right) \\
&\quad + q^3 f\left(-q^{22}, -q^{10}\right) f\left(-q^{30}, -q^2\right) \\
&\quad + q^{10} f\left(-q^{30}, -q^2\right) f\left(-q^{38}, -q^{-6}\right).
\end{aligned}$$

We make the same argument for the product $\psi(-q^4)\psi(-q)$, to find that

$$\psi(-q^4)\psi(-q) = \sum_{a=-1}^{2} (-1)^a q^{2a^2+a}$$
$$\times \sum_{M=-\infty}^{\infty} (-1)^M q^{16M^2+(-2+8a)M}$$
$$\times \sum_{N=-\infty}^{\infty} (-1)^N q^{16N^2+(6+8a)N}$$
$$= \sum_{a=-1}^{2} (-1)^a q^{2a^2+a} f\left(-q^{14+8a},-q^{18-8a}\right)$$
$$\times f\left(-q^{22+8a},-q^{10-8a}\right)$$
$$= -qf\left(-q^6,-q^{26}\right)f\left(-q^{14},-q^{18}\right)$$
$$+f\left(-q^{14},-q^{18}\right)f\left(-q^{22},-q^{10}\right)$$
$$-q^3 f\left(-q^{22},-q^{10}\right)f\left(-q^{30},-q^{2}\right)$$
$$+q^{10} f\left(-q^{30},-q^{2}\right)f\left(-q^{38},-q^{-6}\right).$$

Using the above two identities, we deduce the desired results.

APPLICATIONS TO THE THEORY OF PARTITIONS

Some of our modular relations yield theorems in the theory of partitions. In this section, we present partition theoretic interpretations of some of our modular relations.

Definition 5.1 *A positive integer n has k colors if there are k copies of n available and all of them are viewed as distinct objects. Partitions of positive integer into parts with colors are called "colored partitions".*

For example, if 1 is allowed to have two colors, say r (red) and g (green), then all the colored partitions of 3 are 3, $2+1_r, 2+1_g, 1_r+1_r+1_r, 1_r+1_r+1_g$, $1_r+1_g+1_g$ and $1_g+1_g+1_g$. It is easy to see that

$$\frac{1}{\left(q^u;q^v\right)_\infty^k}$$

TABLE 1. Example for Theorem 5.2

$p_1(6)=8$	$p_2(6)=9$	$p_3(4)=1$
5_r+1	6_r	$2+2$
5_g+1	6_g	
3_r+3_r	$3+3$	
3_r+3_g	$3+2+1$	
3_g+3_g	$3+1+1+1$	
$3_r+1+1+1$	$2+2+2$	
$3_g+1+1+1$	$2+2+1+1$	
$1+1+1+1+1+1$	$2+1+1+1+1$	
	$1+1+1+1+1+1$	

is the generating function for the number of partitions of n where all the parts are congruent to $u \pmod{v}$ and have k colors. For simplicity, we define

$$(q^{r\pm};q^s)_\infty := (q^r, q^{s-r};q^s)_\infty,$$

where r and s are positive integers with $r<s$ and $(a_1,a_2,\ldots,a_n;q)_\infty := \prod_{i=1}^{n}(a_i;q)_\infty$.

Theorem 5.2 *Let* $p_1(n)$ *denote the number of partitions of* n *into parts congruent to,* ± 1, ± 3, ± 5, $\pm 7 \pmod{16}$ *where the parts congruent to* ± 3, $\pm 5 \pmod{16}$ *have two colors. Let* $p_2(n)$ *denote the number of partitions of* n *into parts congruent to* ± 1, ± 2, ± 3, ± 6, $8 \pmod{16}$ *where the parts congruent to* ± 6, $8 \pmod{16}$ *have two colors. Let* $p_3(n)$ *denote the number of partitions of* n *into parts congruent to* ± 2, ± 5, ± 6, ± 7, $8 \pmod{16}$ *where the parts congruent to* ± 6, $8 \pmod{16}$ *have two colors. Then, for any positive integer* $n \geq 2$,

$$p_1(n) = p_2(n) - p_3(n-2)$$

Proof. Using (2), one can easily verify that (32) is equivalent to

TABLE 2. Example for Theorem 5.4

$p_1(4)=7$	$p_2(4)=6$	$p_3(3)=1$
$3+1_r$	2_r+2_r	3
$3+1_g$	2_r+2_g	
$1_r+1_r+1_r+1_r$	2_g+2_g	
$1_r+1_r+1_r+1_g$	2_r+1+1	
$1_r+1_r+1_g+1_g$	2_g+1+1	
$1_r+1_g+1_g+1_g$	$1+1+1+1$	
$1_g+1_g+1_g+1_g$		

$$\begin{aligned}&\frac{1}{(q^{1\pm},q^{3\pm},q^{3\pm},q^{5\pm},q^{5\pm},q^{7\pm};q^{16})_\infty}\\&=\frac{1}{(q^{1\pm},q^{2\pm},q^{3\pm},q^{6\pm},q^{6\pm},q^{8},q^{8};q^{16})_\infty}\\&-\frac{q^2}{(q^{2\pm},q^{5\pm},q^{6\pm},q^{6\pm},q^{7\pm},q^{8},q^{8};q^{16})_\infty}\end{aligned} \tag{56}$$

The three quotients of (56) represent the generating functions for $p_1(n)$, $p_2(n)$ and $p_3(n)$, respectively. Hence, (56) is equivalent to

$$\sum_{n=0}^{\infty} p_1(n)q^n = \sum_{n=0}^{\infty} p_2(n)q^n - q^2\sum_{n=0}^{\infty} p_3(n)q^n,$$

where we set $p_1(0)=p_2(0)=p_3(0)=1$. Equating coefficients of q^n $(n \geq 2)$ on both sides yields the desired result.

Example 5.3 *The following table illustrates the case* $n=6$ *in Theorem 5.2*

Theorem 5.4 *Let* $p_1(n)$ *denote the number of partitions of* n *into parts congruent to* ± 1, ± 3, ± 5, $\pm 7 \pmod{16}$ *where the parts congruent to* ± 1, $\pm 7 \pmod{16}$ *have two colors. Let* $p_2(n)$ *denote the number of partitions of* n *into parts con-*

gruent to ±1, ±2, ±5, ±6, $8 \pmod{16}$ *where the parts congruent to* ±2, $8 \pmod{16}$ *have two colors. Let* $p_3(n)$ *denote the number of partitions of* n *into parts congruent to* ±2, ±3, ±6, ±7, $8 \pmod{16}$ *where the parts congruent to* ±2, $8 \pmod{16}$ *have two colors. Then, for any positive integer* $n \geq 1$,

$$p_1(n) = p_2(n) - p_3(n-1).$$

Proof. Using (2), one can easily verify that (33) is equivalent to

$$\begin{aligned}&\frac{1}{(q^{1\pm},q^{1\pm},q^{3\pm},q^{5\pm},q^{7\pm},q^{7\pm};q^{16})_\infty}\\&=\frac{1}{(q^{1\pm},q^{2\pm},q^{2\pm},q^{5\pm},q^{6\pm},q^{8},q^{8};q^{16})_\infty}\\&+\frac{q}{(q^{2\pm},q^{2\pm},q^{3\pm},q^{6\pm},q^{7\pm},q^{8},q^{8};q^{16})_\infty}\end{aligned} \tag{57}$$

The three quotients of (57) represent the generating functions for $p_1(n)$, $p_2(n)$ and $p_3(n)$, respectively. Hence, (57) is equivalent to

$$\sum_{n=0}^{\infty} p_1(n)q^n = \sum_{n=0}^{\infty} p_2(n)q^n + q\sum_{n=0}^{\infty} p_3(n)q^n,$$

where we set $p_1(0) = p_2(0) = p_3(0) = 1$. Equating coefficients of q^n $(n \geq 1)$ on both sides yields the desired result.

Example 5.5 *The following table illustrates the case* $n=4$ *in Theorem 5.4*

In a similar way one can prove the following four theorems using the modular relations (14), (16), (18) and (19), respectively:

Theorem 5.6 *Let* $p_1(n)$ *denote the number of partitions of* n *into parts not congruent to* ±7, ±9, ±16, ±21, $\pm23 \pmod{48}$ *where the parts congruent to* ±12, $24 \pmod{48}$ *have two colors and* ±6, $\pm18 \pmod{48}$ *have three colors. Let* $p_2(n)$ *denote the number of partitions of* n *into parts not congruent to* ±5, ±11, ±15, ±16, $\pm21 \pmod{48}$ *where the parts congruent to* $\pm12, 24 \pmod{48}$ *have two colors and* ±6, $\pm18 \pmod{48}$ *have three colors. Let* $p_3(n)$ *denote the number of partitions of* n *into parts not congruent to* ±3, ±9, ±13, ±16, $\pm19 \pmod{48}$ *where the parts congruent to* ±12, $24 \pmod{48}$ *have two colors and* ±6, $\pm18 \pmod{48}$ *have three colors. Let* $p_4(n)$ *denote the number of partitions of* n *into parts not congruent to* ±1, ±3, ±15, ±16, $\pm17 \pmod{48}$ *where the parts congruent to* $\pm12, 24 \pmod{48}$

have two colors and ± 6, $\pm 18 \pmod{48}$ *have three colors. Let* $p_5(n)$ *denote the number of partitions of* n *into odd parts with two colors. Then, for any positive integer* $n \geq 6$,

$$p_1(n) + p_2(n-1) + p_3(n-3) + p_4(n-6) = p_5(n).$$

Theorem 5.7 *Let* $p_1(n)$ *denote the number of partitions of* n *into parts congruent to* ± 2, ± 4, ± 6, ± 8, ± 10, $16 \pmod{32}$ *where the parts congruent to* ± 2, ± 6, ± 10, $16 \pmod{32}$ *have two colors and* $\pm 8 \pmod{32}$ *have three colors. Let* $p_2(n)$ *denote the number of partitions of* n *into parts congruent to* ± 2, ± 4, ± 6, ± 8, ± 14, $16 \pmod{32}$ *where the parts congruent to* ± 2, ± 6, ± 14, $16 \pmod{32}$ *have two colors and* $\pm 8 \pmod{32}$ *have three colors. Let* $p_3(n)$ *denote the number of partitions of* n *into parts congruent to where the parts congruent to* ± 2, ± 4, ± 8, ± 10, ± 14, $16 \pmod{32}$ *have two colors and* $\pm 8 \pmod{32}$ *have three colors. Let* $p_4(n)$ *denote the number of partitions of* n *into parts congruent to* ± 2, ± 10, ± 14, $16 \pmod{32}$ *where the parts congruent to* ± 4, ± 6, ± 8, ± 10 ± 14, $16 \pmod{32}$ *have two colors and* $\pm 8 \pmod{32}$ *have three colors. Let* $p_5(n)$ *denote the number of partitions of* n *into parts congruent to* ± 1, ± 2, ± 3, ± 5, ± 6, ± 7, ± 9, ± 10, ± 11, ± 13, ± 14, $\pm 15 \pmod{32}$. *Then, for any positive integer* $n \geq 6$,

$$p_1(n) + p_2(n-1) + p_3(n-3) + p_4(n-6) = p_5(n).$$

Theorem 5.8 *Let* $p_1(n)$ *denote the number of partitions of* n *into parts congruent to* ± 3, ± 5, ± 7, ± 9, ± 11, ± 13, ± 19, $\pm 23 \pmod{48}$. *Let* $p_2(n)$ *denote the number of partitions of* n *into parts congruent to* ± 1, ± 5, ± 7, ± 9, ± 11, ± 17, ± 21, $\pm 23 \pmod{48}$ *Let* $p_3(n)$ *denote the number of partitions of* n *into parts congruent to* ± 1, ± 3, ± 7, ± 13, ± 15, ± 17, ± 19, $\pm 23 \pmod{48}$ *Let* $p_4(n)$ *denote the number of partitions of* n *into parts congruent to* ± 1, ± 5, ± 11, ± 13, ± 15, ± 17, ± 19, $\pm 21 \pmod{48}$. *Then, for any positive integer* $n \geq 3$,

$$p_1(n) + p_3(n-1) = p_2(n) + p_4(n-3).$$

Theorem 5.9 *Let* $p_1(n)$ *denote the number of partitions of* n *into parts not congruent to* ± 2, ± 5, ± 6, ± 10, ± 11, ± 14, ± 16, ± 18, ± 21, $\pm 22 \pmod{48}$ *where the parts congruent to* ± 3, ± 9, ± 15, $24 \pmod{48}$ *have two colors,* ± 4, $\pm 20 \pmod{48}$ *have three colors and* $\pm 12 \pmod{48}$ *have four colors. Let* $p_2(n)$ *denote the number of partitions of* n *into parts not congruent to* ± 1, ± 2, ± 6, ± 10, ± 14, ± 15, ± 16, ± 17, ± 18, $\pm 22 \pmod{48}$, *where the parts congruent*

$\pm3,\ \pm9,\ \pm21,\ 24 \pmod{48}$ *to, have two colors* $\pm4,\ \pm20 \pmod{48}$ *have three colors and* $\pm12 \pmod{48}$ *have four colors. Let* $p_3(n)$ *denote the number of partitions of into parts not congruent to,* $\pm2, \pm6,\ \pm7,\ \pm9,\ \pm10,\ \pm14,\ \pm16,\ \pm18,\ \pm22,\ \pm23 \pmod{48}$ *where the parts congruent to* ±3 , ±15 , ±21 , $24 \pmod{48}$ *have two* $\pm4,\ \pm20 \pmod{48}$ *colors, have three colors* $\pm12 \pmod{48}$ *and have four colors. Let* $p_3(n)$ *denote the number of partitions of into parts not congruent to* $\pm2,\ \pm3,\ \pm6,\ \pm10,\ \pm13,\ \pm14, \pm16,\ \pm18, \pm19,\ \pm22 \pmod{48}$ *where the parts congruent to* $\pm9, \pm15$, ±21 , $24 \pmod{48}$ *have two colors,* ±4 , $\pm20 \pmod{48}$ *have three colors* $\pm12 \pmod{48}$ *and have four colors. Let* $p_5(n)$ *denote the number of partitions of into parts not congruent to* $\pm4,\ \pm8,\ \pm12,\ \pm16,\ \pm20,\ 24 \pmod{48}$ *where the parts congruent to* $\pm2,\ \pm6,\ \pm10,\ \pm14,\ \pm18,\ \pm22 \pmod{48}$ *have two colors. Then, for any positive integer,* $n \geq 5$,

$$\begin{aligned} &p_1(n) + p_2(n-2) + p_3(n-2) \\ &\qquad = p_4(n-5) + p_5(n). \end{aligned}$$

REFERENCES

1. C. Adiga, B. C. Berndt, S. Bhargava and G. N. Watson, Chapter 16 of Ramanujan's second notebook: Theta functions and q -series, Mem. Amer. Math. Soc. 315 (1985), 1-91.
2. C. Adiga and N. A. S. Bulkhali, Some modular relations analogues to the Ramanujan's forty identities with its applications to partitions, Axioms, 2(1) (2013), 20-43.
3. C. Adiga and N. A. S. Bulkhali, On certain new modular relations for the Rogers-Ramanujan type functions of order ten and its applications to partitions, Note di Matematica, to appear.
4. C. Adiga and M. S. Surekha, Some modular relations for the Roger-Ramanujan type functions of order six and its applications to partitions, Proc. J. math. Soc., to appear.
5. C. Adiga and M. S. Surekha and N. A. S. Bulkhali, Some modular relations of order six with its applications to partitions, Gulf J. Math., to appear.
6. C. Adiga and A. Vanitha, New modular relations for the Rogersâ€"Ramanujan type functions of order fifteen,Notes on Numb. Theor. Discrete Math., 20(1) (2014), 36-48.
7. C. Adiga and A. Vanitha and N. A. S. Bulkhali, Modular relations for the Rogers-Ramanujan-Slater type functions of order fifteen and its applications to partitions, Romanian J. Math. Comput. Sci., 3(2) (2013), 119-139.

8. C. Adiga and A. Vanitha and N. A. S. Bulkhali, Some modular relations for the Rogers-Ramanujan type functions of order fifteen and its applications to partitions, Palestine J. Math., 3(2) (2014), 204-217.
9. C. Adiga, K. R. Vasuki and N. Bhaskar, Some new modular relations for the cubic functions, South East Asian Bull. Math., 36 (2012), 1-19.
10. C. Adiga, K. R. Vasuki and B. R. Srivatsa Kumar, On modular relations for the functions analogous to Rogers-Ramanujan functions with applications to partitions, South East J. Math. and Math. Sc., 6(2) (2008), 131-144.
11. G. E. Andrews and B. C. Berndt, Ramanujans Lost Notebook, Part III, Springer, New York, 2012.
12. N. D. Baruah and J. Bora, Further analogues of the Rogers-Ramanujan functions with applications to partitions,Elec. J. Combin. Number Thy., 7(2) (2007), # A05, 22pp.
13. N. D. Baruah and J. Bora, Modular relations for the nonic analogues of the Rogers-Ramanujan functions with applications to partitions, J. Number Thy., 128 (2008), 175-206.
14. N. D. Baruah, J. Bora and N. Saikia, Some new proofs of modular relations for the Göllnitz-Gordon functions,Ramanujan J., 15 (2008), 281-301.
15. B. C. Berndt and H. Yesilyurt, New identities for the Rogers-Ramanujan function, Acta Arith., 120 (2005), 395-413.
16. B. C. Berndt, G. Choi, Y. S. Choi, H. Hahn, B. P. Yeap, A. J. Yee, H. Yesilyurt, and J. Yi, Ramanujan's forty identities for the Rogers-Ramanujan function, Mem., Amer. Math. Soc., 188(880) (2007), 1-96.
17. A. J. F. Biagioli, A proof of some identities of Ramanujan using modular functions, Glasg. Math. J., 31 (1989), 271-295.
18. B. J. Birch, A look back at Ramanujan's Notebooks, Math. Proc. Camb. Soc., 78 (1975), 73-79.
19. R. Blecksmith. J. Brillhat. and I. Gerst. A foundamental modular identity and some applications. Math. Comp., 61 (1993), 83-95.
20. D. Bressoud, Proof and Generalization of Certain Identities Conjectured by Ramanujan, Ph.D. Thesis, Temple University, 1977.
21. D. Bressoud, Some identities involving Rogers-Ramanujan-type functions, J. London Math. Soc., 16(2) (1977), 9-18.
22. S. L. Chen and S.-S. Huang, New modular relations for the Göllnitz-Gordon functions, J. Number Thy., 93 (2002), 58-75.
23. C. Gugg, Modular Identities for the Rogers-Ramanujan Function and Analogues, Ph.D. Thesis, University of Illinois at Urbana-Champaign, 2010.
24. H. Hahn, Septic analogues of the Rogers-Ramanujan functions, Acta Arith., 110 (4) (2003), 381-399.

25. H. Hahn, Eisenstein Series, Analogues of the Rogers-Ramanujan Functions, and Parttition Identities, Ph.D. Thesis, University of Illinois at Urbana-Champaign, 2004.

26. S.-S. Huang, On modular relations for the Göllnitz-Gordon functions with applications to partitions, J. Number Thy., 68 (1998), 178-216.

27. S. Ramanujan, Algebraic relations between certain infinite products, Proc. London Math. Soc., 2 (1920), p. xviii.

28. S. Ramanujan, The Lost Notebook and Other Unpublished Papers, Narosa, New Delhi, 1988.

29. S. Robins, Arithmetic Properties of Modular Forms, Ph.D. Thesis, University of California at Los Angeles, 1991.

30. L. J. Rogers, On a type of modular relation, Proc. London Math. Soc., 19 (1921), 387-397.

31. L. J. Slater, Further identities of Rogers-Ramanujan type, London Math. Soc., 54(2) (1952), 147-167.

32. K. R. Vasuki and P. S. Guruprasad, On certain new modular relations for the Rogers-Ramanujan type functions of order twelve, Adv. Stud. Contem. Math., 20(3) (2000), 319-333.

33. K. R. Vasuki, G. Sharath, and K. R. Rajanna, Two modular equations for squares of the cubic-functions with applications, Note di Matematica, 30(2) (2010), 61-71.

34. G. N. Watson, Proof of certain identities in combinatory analysis, J. Indian Math. Soc., 20 (1933), 57-69.

35. E. X. W. Xia and X. M. Yao, Some modular relations for the Göllnitz-Gordon functions by an even-odd method,J. Math. Anal. Appl., 387 (2012), 126-138.

36. H. Yesilyurt, A generalization of a modular identity of Rogers, J. Number Thy., 129 (2009), 1256-1271.

17

ON INEQUALITIES OF TRIGONOMETRICALLY P- CONVEX FUNCTIONS

Mohamed Sabri Salem Ali[1]

[1] Department of Mathematics, Faculty of Education, Ain Shams University, Cairo, Egypt

ABSTRACT

The main goal of this paper is to derive two integral inequalities for trigonometrically ρ- convex functions which are closely connected with Andersson's inequality for ordinary convex functions.

INTRODUCTION AND PRELIMINARIES

Trigonometrically ρ - convex functions play a central role in many various aspects in the theory of entire functions (of order $0 < \rho < \infty$) and in the theory of cavitational diagrams for hydroprofiles, see for example [4,5,6]. In what follows, we shall be concerned with real finite functions defined on a finite or infinite interval $I \subset \mathbb{R}$. In this section we present the basic definitions and results which will be used later, for more information see [5,6]

Definition 1.1. A function $f : I \to \mathbb{R}$ is said to be trigonometrically ρ - convex, if for any arbitrary closed subinterval $[u,v]$ of I such that $0 < \rho(v - u) < \pi$, the graph of $f(x)$ for $x \in [u, v]$ lies nowhere above the ρ - trigonometric function, determined by the equation:

$$H(x) = H(x; u, v, f) = A \cos \rho x + B \sin \rho x$$

where A and B are chosen such that

$$H(u)=f(u), \text{ and } H(v)=f(v)$$

Equivalently, if for all $x \in [u, v]$

$$f(x) \le H(x) = \frac{f(u)\sin\rho(v-x)+f(v)\sin\rho(x-u)}{\sin\rho(v-u)}$$

Definition 1.2. A function

$$T_u(x) = A\cos\rho x + B\sin\rho x$$

is said to be supporting function for $f(x)$ at the point $u \in I$, if

$$T_u(u) = f(u), \quad and \quad T_u(x) \le f(x) \quad \forall x \in I$$

That is, if $f(x)$ and $T_u(x)$ agree at $x = u$, and the graph of $f(x)$ does not lie under the support curve

Theorem 1.1. [1] A function $f: I \to \mathbb{R}$ is trigonometrically ρ - convex on I if and only if there exists a supporting function for $f(x)$ at each point $x \in I$.

Property 1.1.[1] If $f: I \to \mathbb{R}$ is differentiable trigonometrically ρ - convex function, then the supporting function for $f(x)$ at the point $u \in I$ has the form:

$$T_u(x) = f(u)\cos\rho(x-u) + f'(u)\sin\rho(x-u).$$

To keep the paper self contained we present the following results [7].

Theorem 1.2. If $f: (a,b) \to \mathbb{R}$ is differentiable function. Then f is convex if and only if f' is increasing.

Theorem 1.3. A function $f: (a,b) \to \mathbb{R}$ is convex if and only if there is an increasing function $g: (a,b) \to \mathbb{R}$ and a point $c \in (a,b)$ such that for all $x \in (a,b)$,

$$f(x) - f(c) = \int_c^x g(t)\,dt$$

Theorem 1.4. If $f:(a,b)\to\mathbb{R}$ and $g:(a,b)\to\mathbb{R}$ are both non-negative, increasing (decreasing), and convex, then $h(x)=f(x)g(x)$ also preserves these three properties.

In [3], B. J. Andersson established the following theorem

Theorem 1.5. Let $F_1(x), F_2(x), \dots, F_n(x)$ be convex functions, defined in $0 \le x \le 1$, and for which

$$F_p(x) \ge 0, \quad F_p(0) = 0, \quad p = 1, 2, \dots, n$$

If $\int_0^1 F_p(x)dx = \alpha_p$, then

$$\int_0^1 F_1(x)F_2(x)\dots F_n(x)dx \ge \frac{2^n}{n+1}\alpha_1\ \alpha_2 \dots \alpha_n. \tag{1}$$

RESULTS

In this section we derive a similar result to Andersson's inequality for trigonometrically ρ - convex functions. For more inequalities, one may refer to [2March, 2013."].

Theorem 2.1. Let $g_1(x), g_2(x), \dots, g_n(x)$ be continuous convex functions, defined in $0 \le x \le \frac{\pi}{2\rho}$, for which

$$g_k(x) \ge 0, \quad g_k(0) = 0, \quad k = 1, 2, \dots, n$$

and let $f(x)$ be a non-negative, $\frac{2\pi}{\rho}$ - periodic, differentiable, and trigonometrically ρ - convex function defined on $\mathbb{R}$, such that: $f'(0)=0$ and

$$\int_0^{\frac{\pi}{2\rho}} g_k(x)\cos\rho x\ dx = \alpha_k,$$

then

$$\int_0^{\frac{\pi}{2\rho}} f(x)\prod_{k=1}^{n} g_k(x)\,dx \ge \frac{2^n f(0)}{\rho(n+1)}\prod_{k=1}^{n}\alpha_k$$

Proof. As $f(x)$ is trigonometrically ρ- convex function, then from Definition 1.2, it follows that:

$$f(x) \geq T_0(x) \qquad \forall x \in [0, \frac{\pi}{2\rho}]$$

Since $f(x)$ is differentiable and $f'(0) = 0$, then from Property 1.1, the supporting function $T_0(x)$ for $f(x)$ at the point $0 \in [0, \frac{\pi}{2\rho}]$ can be written in the form

$$T_0(x) = f(0)\cos \rho x$$

Consequently,

$$f(x) \geq f(0)\cos \rho x \qquad \forall x \in [0, \frac{\pi}{2\rho}]. \tag{2}$$

As $\prod_{k=1}^{n} g_k(x) \geq 0$, by using (2), one has:

$$\int_0^{\frac{\pi}{2\rho}} f(x) \prod_{k=1}^{n} g_k(x)\, dx \geq f(0) \int_0^{\frac{\pi}{2\rho}} \prod_{k=1}^{n} g_k(x) \cos \rho x \, dx. \tag{3}$$

Using the following substitution

$$t = \sin \rho x, \tag{4}$$

and let

$$h_k(t) = g_k(x), \tag{5}$$

then it follows that:

$$h_k(0) = 0, \quad h_k(t) \geq 0 \qquad \forall t \in [0,1],$$

and $h_k'(t)$ is an increasing function in $[0,1]$. Using Theorem 1.2, one obtains that $h_k(t)$ is a convex function in $[0,1]$.

Let

$$\int_0^1 h_k(t)\,dt = \alpha_k, \quad k = 1,2,\dots,n$$

we obseve that the functions $h_k(t)$, $k = 1,2,\dots,n$ satisfy all assumptions of the Theorem 1.5 in the interval.[0,1]

Hence, using (1), one obtains:

$$\int_0^1 \prod_{k=1}^n h_k(t)\,dt \geq \frac{2^n}{n+1}\prod_{k=1}^n \alpha_k. \tag{6}$$

Now using (4), (5), and (6), then (3) turns out to:

$$\int_0^{\frac{\pi}{2\rho}} f(x)\prod_{k=1}^n g_k(x)dx \geq \frac{f(0)}{\rho}\int_0^1 \prod_{k=1}^n h_k(t)\,dt$$
$$\geq \frac{2^n f(0)}{\rho(n+1)}\prod_{k=1}^n \alpha_k,$$

where $\alpha_k = \int_0^1 h_k(t)dt = \int_0^{\frac{\pi}{2\rho}} g_k(x)\cos\rho x\;dx$.

Hence, the claim.

Now, we are in a position to prove the following main theorem.

Theorem 2.2. Let $g_1(x), g_2(x), \dots, g_n(x)$ be convex functions, defined in $0 \leq x \leq \frac{\pi}{2\rho}$, for which

$$g_r(x) \geq 0, \quad g_r(0) = 0, \quad r = 1,2,\dots,n$$

and let $f(x)$ be a non-negative, $\frac{2\pi}{\rho}$- periodic, differentiable, and trigonometrically ρ- convex function defined on $\mathbb{R}$, such that: $f'(\pi/2\rho) = 0$ and $\int_0^{\frac{\pi}{2\rho}} g_r(x)\,dx = \alpha_r$.

Then, one has the following sharp inequality:

$$\int_0^{\frac{\pi}{2\rho}} f(x)\prod_{r=1}^{n} g_r(x)\,dx \geq \frac{\pi}{(n+2)} f(\pi/2\rho)\left(\frac{4\rho}{\pi}\right)^{n-1} \times \left(\prod_{r=1}^{n}\alpha_r\right){}_1F_2(1+\frac{n}{2};\frac{3}{2},2+\frac{n}{2};-\frac{\pi^2}{16}). \tag{7}$$

Proof. The proof will be divided into 4 steps.

Let M denote the class of convex functions of the theorem.

Step 1. If $g \in M$, then g is increasing.

Since $g(x)$ is convex in $[0,\frac{\pi}{2\rho}]$ and $g(0)=0$, then from Theorem 1.3 there is an increasing function $h:[0,\frac{\pi}{2\rho}]\rightarrow\mathbb{R}$

$$g(x)=\int_0^x h(t)dt, \quad x\in[0,\frac{\pi}{2\rho}]. \tag{8}$$

Now, suppose that $h(t_0)<0$ for some $t_0\in[0,\frac{\pi}{2\rho}]$.

As h is increasing, then $h(t)<0$ for all $t\in[0,t_0]$, therefore, $\int_0^x h(t)dt<0, \quad x\in[0,t_0]$.

From (8) it follows, $g(x)<0, \quad x\in[0,t_0]$, which contradicts the fact that $g(x)\geq 0, \quad x\in[0,\frac{\pi}{2\rho}]$.

Thus, $h(t)\geq 0$ for all $t\in[0,\frac{\pi}{2\rho}]$.

Now, let $x_1, x_2\in[0,\frac{\pi}{2\rho}]$, if $x_1<x_2$, then using (8), one has:

$$0\leq \int_{x_1}^{x_2} h(t)dt = \int_0^{x_2} h(t)dt - \int_0^{x_1} h(t)dt = g(x_2)-g(x_1)$$

Hence, $g(x)$ is an increasing function.

For the next steps, let

$$g_r^*(x)=\frac{2\alpha_r x}{(\pi/2\rho)^2}, \qquad 0\leq x\leq\frac{\pi}{2\rho}, \tag{9}$$

And

$$\Phi_r(x)=\int_0^x\left[g_r^*(s)-g_r(s)\right]ds. \tag{10}$$

Step 2. M is closed under multiplication.

From (9), it follows that:

$g_r^*(x)$ is non-negative, increasing, and convex function satisfies $g_r^*(0)=0$.

Thus, $g_r^*\in M$, $r=1,2,\dots,n$.

Hence, from Theorem 1.4, M is closed under multiplication.

This confirm that this property is independent of the choice of g_r's or g_r^*'s.

Step 3. $\Phi_r(x)\geq 0,\quad x\in[0,\frac{\pi}{2\rho}]$.

Using (9) and (10), one concludes that

$$\int_0^{\frac{\pi}{2\rho}} g_r^*(x)\,dx=\int_0^{\frac{\pi}{2\rho}} g_r(x)\,dx=\alpha_r, \tag{11}$$

$$\Phi_r(0)=0,\ \text{and}\ \Phi_r(\pi/2\rho)=0. \tag{12}$$

From the convexity of $g_r(x)$, it follows that the graph of $g_r(x)$, must intersect the straight line of $g_r^*(x)$ in a unique point q as shown in Figure 1.

If x lies in [0,p], then obviously from (10) it followos that

$$\Phi_r(x)\geq 0,\qquad 0\leq x\leq p$$

Otherwise, if $p\leq x\leq\frac{\pi}{2\rho}$, one has:

$$\int_x^{\frac{\pi}{2\rho}} g_r(s)ds\geq\int_x^{\frac{\pi}{2\rho}} g_r^*(s)ds$$

Using (11), one obtains:

$$\int_0^{\frac{\pi}{2\rho}} g_r^*(s)ds - \int_x^{\frac{\pi}{2\rho}} g_r^*(s)ds \geq \int_0^{\frac{\pi}{2\rho}} g_r(s)ds - \int_x^{\frac{\pi}{2\rho}} g_r(s)ds$$

$$\int_0^x \left[g_r^*(s) - g_r(s)\right]ds \geq 0$$

Thus, $\Phi_r(x) \geq 0, \quad p \leq x \leq \frac{\pi}{2\rho}$.

Hence $\Phi_r(x) \geq 0, \quad x \in [0, \frac{\pi}{2\rho}]$.

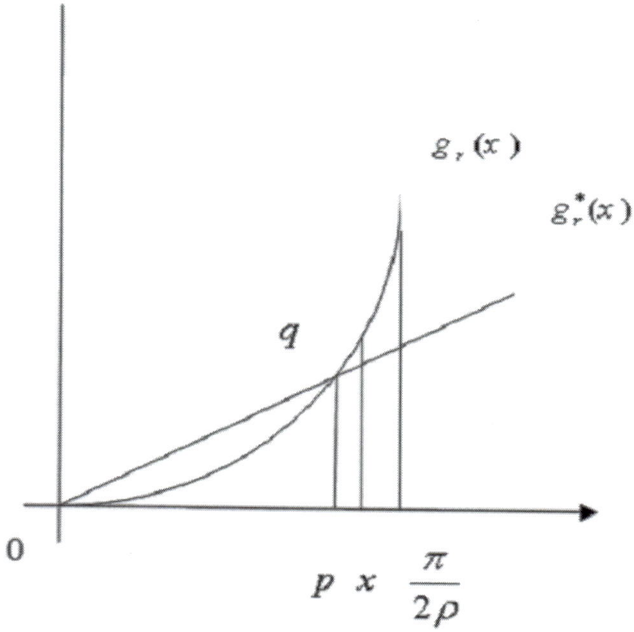

Figure 1.

As $f(x)$ is differentiable trigonometrically ρ- convex function and $f'(\pi/2\rho) = 0$, according to Property 1.1, for convenience, we denote

$$f^*(x) = f(\pi/2\rho)\sin\rho x, \tag{13}$$

as the supporting function for $f(x)$ at the point $\frac{\pi}{2\rho} \in [0, \frac{\pi}{2\rho}]$.

But, from Definition 1.2, one obtains:

$$f(x) \geq f^*(x), \qquad x \in [0, \frac{\pi}{2\rho}], \tag{14}$$

hence we can go to:

Step 4. We show that If $G_1(x), G_2(x) \in M$, then

$$\int_0^{\frac{\pi}{2\rho}} G_1(x)G_2(x)f(x)dx \geq \int_0^{\frac{\pi}{2\rho}} G_1^*(x)G_2(x)f^*(x)dx. \tag{15}$$

From Step 2, we observe:

$$G_1(x)G_2(x) \geq 0, \qquad x \in [0, \frac{\pi}{2\rho}]$$

Using (14), one has,

$$\int_0^{\frac{\pi}{2\rho}} G_1(x)G_2(x)f(x)dx \geq \int_0^{\frac{\pi}{2\rho}} G_1(x)G_2(x)f^*(x)dx. \tag{16}$$

Let

$$S_1(x) = G_2(x)f^*(x). \tag{17}$$

Using (10) and (12), it follows that

$$\int_0^{\frac{\pi}{2\rho}} G_1(x)S_1(x)dx = \int_0^{\frac{\pi}{2\rho}} S_1(x)[G_1^* - (G_1^* - G_1)](x)dx$$

$$= \int_0^{\frac{\pi}{2\rho}} S_1(x)G_1^*(x)dx - \int_0^{\frac{\pi}{2\rho}} S_1(x)\, d\Phi_1(x)$$

$$= \int_0^{\frac{\pi}{2\rho}} G_1^*(x)S_1(x)dx + \int_0^{\frac{\pi}{2\rho}} \Phi_1(x)\, dS_1(x).$$

Since $dS_1(x) \geq 0$, and from Step 3, we infer that

$$\int_0^{\frac{\pi}{2\rho}} \Phi_1(x)\,dS_1(x) \geq 0.$$

Thus

$$\int_0^{\frac{\pi}{2\rho}} G_1(x)S_1(x)\,dx \geq \int_0^{\frac{\pi}{2\rho}} G_1^*(x)S_1(x)\,dx. \tag{18}$$

Hence, from (16), (17) and (18), we obtain the required inequality (15).

Now, we prove the main inequality (7).

From Step 2, we have

$$g_n(x),\ g_1(x)g_2(x)\ldots g_{n-1}(x) \in M$$

Thus, using (15), one has

$$\int_0^{\frac{\pi}{2\rho}} g_1(x)\ldots g_{n-1}(x)g_n(x)f(x)\,dx \geq$$

$$\int_0^{\frac{\pi}{2\rho}} g_1(x)\ldots g_{n-1}(x)g_n^*(x)f^*(x)\,dx.$$

Again, $g_{n-1}(x),\ g_1(x)\ldots g_{n-2}(x)g_n^*(x) \in M$.

Hence, from (18) it follows that

$$\int_0^{\frac{\pi}{2\rho}} g_1(x)\ldots g_{n-1}(x)g_n^*(x)f^*(x)\,dx \geq$$

$$\int_0^{\frac{\pi}{2\rho}} g_1(x)\ldots g_{n-1}^*(x)g_n^*(x)f^*(x)\,dx.$$

Repeating the above argument and using (18) each time, then from (9) and (13) one obtains:

$$\int_0^{\frac{\pi}{2\rho}} g_1(x)\dots g_n(x)f(x)dx \geq \int_0^{\frac{\pi}{2\rho}} g_1^*(x)\dots g_n^*(x)f^*(x)dx$$

$$= \frac{2^n f(\pi/2\rho)}{(\pi/2\rho)^{2n}}\left(\prod_{r=1}^{n}\alpha_r\right)\int_0^{\frac{\pi}{2\rho}} x^n \sin\rho x\, dx$$

Using the following substitution

$z = \rho x$,

we get

$$\int_0^{\frac{\pi}{2\rho}} f(x)\prod_{r=1}^{n} g_r(x)dx \geq$$

$$\rho^n\left(\frac{8\rho}{\pi^2}\right)^n f(\pi/2\rho)\left(\prod_{r=1}^{n}\alpha_r\right)\frac{1}{\rho^{n+1}}\int_0^{\frac{\pi}{2}} z^n \sin z\, dz$$

$$= \frac{1}{\rho} f(\pi/2\rho)\left(\frac{8\rho}{\pi^2}\right)^n\left(\prod_{r=1}^{n}\alpha_r\right)\times$$

$$\frac{(\pi/2)^{n+2}}{(n+2)}\ {}_1F_2(1+\frac{n}{2};\frac{3}{2},2+\frac{n}{2};-\frac{\pi^2}{16})$$

$$= \frac{\pi}{(n+2)} f(\pi/2\rho)\left(\frac{4\rho}{\pi}\right)^{n-1}\left(\prod_{r=1}^{n}\alpha_r\right)\times$$

$${}_1F_2(1+\frac{n}{2};\frac{3}{2},2+\frac{n}{2};-\frac{\pi^2}{16}),$$

and the theorem is proved.

Note: the equality in (7) occurs if

$$f(x) = f(\pi/2\rho)\sin\rho x,$$

$$g_r(x) = \frac{2\alpha_r x}{(\pi/2\rho)^2}, \qquad r = 1,2,\dots,n.$$

ACKNOWLEDGEMENT

The author wishes to thank the anonymous referees for their fruitful comments and suggestions which improved the original manuscript.

REFERENCES

1. Ali M. S. S., "On Certain Properties of Trigonometrically ρ- Convex Functions," Advances in Pure Mathematics., 2, 337-340, 2012.
2. Ali M. S. S., "On Hadamard's Inequality for Trigonometrically ρ- Convex Functions," accepted to appear inTheoretical Mathematics & Applications., March, 2013.
3. Andersson B. J., "An inequality for Convex Functions," Nordisk Matematisk Tidskrift., 6, 25-26, 1968.
4. Avhadief F. G. and Maklakov D. V., "A Theory of Pressure Envelopes for Hydrofoils," Journal of Ship Resrarch., 42, 81-102, 1995.
5. Levin B. Ya., "Lectures on Entire Functions," American Mathematical Society, 1996.
6. Maergoiz L. S., "Asymptotic Characteristics of Entire Functions and their Applications in Mathematics and Biophysics," Kluwer Academic Publishers, 2003
7. Roberts A. W. and Varberg D. E., "Convex Functions," Academic Press, New York - London, 1973.

18

COMMON AND COINCIDENCE FIXED POINT THEOREMS FOR ASYMPTOTICALLY REGULAR MAPPINGS IN HILBERT SPACES

R.A. Rashwan*

* Department of Mathematics, Assiut University, Assiut, Egypt

ABSTRACT

In this paper we prove common and coincidences fixed point theorems for asymptotically regular mappings under various contractive conditions on a Hilbert space setting. We also study the well – posedness of a common fixed point problem. Our results generalize several well known results in the literature.

INTRODUCTIONS AND PRELIMINARIES

Most of fixed point theorems for mappings in metric spaces satisfying different contraction conditions may be extended to the abstract spaces like Hilbert spaces, Banach spaces and locally convex spaces etc., with some modifications Banach fixed point theorem and its applications are well known. Many authors have extended this theorem, introducing more general contractive conditions, which imply the existence of a fixed point. Almost all of conditions imply the asymptotic regularity of the mappings under considerations. So the investigation of the asymptotically regular maps play an important role in fixed point theory.

Sharma and Yuel [14] and Guay and Singh [7] were among the first who used the concept of asymptotic regularity to prove fixed point theorems for wider class of mappings than a class of mappings introduced and studied by Ćirić [6].

The purpose of this paper is to prove some common and coincidences fixed point theorems in Hilbert spaces and we study the well- posedness of their fixed point problem.

Definition 1.1. [1]. A self mapping T on a closed subset of a Hilbert space H is said to be asymptotically regular at a point x in H, if

$$\left\| T^n x - T^{n+1} x \right\| \rightarrow 0 \ \ as \ \ n \rightarrow \infty,$$

where $T^n x$ denotes the nth iterate of T at x.

Definition 1.2. [1]. Let C be a closed subset of a Hilbert space H. A sequence $\{x_n\}$ in C is said to be asymptotically T - regular if $\|x_n - Tx_n\| \rightarrow 0$ as $n \rightarrow \infty$.

Definition 1.3. [5]. A pair of mappings (f, T) on a Hilbert space H is said to be weakly compatible if f and T commute at their coincidence point (i.e. f T x = T f x whenever f x = T x). A point $y \in H$ is called point of coincidence of two self –mappings f and T on H if there exists a point $y \in H$ such that $y = Tx = fx$.

The following lemma was given in [5] in a metric space setting.

Lemma 1.1. Let X be a non-empty set and the mappings $T, f: X \rightarrow X$ have a unique point of coincidence v in X. If the pair (f, T) is weakly compatible then T and f have a unique common fixed point.

Let H be a Hilbert space,T and f be self –mappings on H with $T(H) \subset f(H)$ and $x_0 \in H$. Choose a point $x_1 \in H$ such that $fx_1 = Tx_0$. This can be done since $T(H) \subset f(H)$.Continuing this process, having chosen $x_1, \ldots, x_n$, we choose x_{n+1} in H such that

$$fx_{n+1} = Tx_n, \quad n = 0,1,2,\ldots.$$

The sequence $\{fx_n\}$ is called a T –sequence with initial point x_0.

Definition1.4. Let T and f be self – mappings on a Hilbert space H, with $T(H) \subset f(H)$ and $x_0 \in H$. A mapping T is said to be asymptotically f – regular at a point x_0 if :

$$\left\| f x_n - f x_{n+1} \right\| \rightarrow 0 \quad as \quad n \rightarrow \infty,$$

Where $\{fx_n\}$ is a T – sequence with initial point x_0.

We know that a Banach space is a Hilbert space iff its norm satisfies the parallelogram law i.e., every $x, y \in H$,

$$\|x+y\|^2+\|x-y\|^2=2(\|x\|^2+\|y\|^2), \tag{1.1}$$

which implies,

$$\|x+y\|^2 \le 2\,(\|x\|^2+\|y\|^2) \tag{1.2}$$

COMMON FIXED POINT THEOREMS

S. T. Patel and at.al. [9] gave the following theorem:

Theorem 2.1. Let C be a closed subset of a Hilbert space H and T be a mapping on H into it self satisfying

$$\begin{aligned}\|Tx-Ty\|^2 \le\ & \alpha\ \|x-Tx\|^2+\beta\ \|y-Ty\|^2 \\ & +\gamma\ \|x-y\|^2+\delta \min\left\{\|x-Ty\|^2, \|y-Tx\|^2\right\} \\ & +\eta \frac{\|y-Tx\|^2}{1+\|x-Tx\|\|x-Ty\|},\end{aligned} \tag{2.1}$$

for all x, y in H, where α, β, γ *and* δ are non –negative reals with $\alpha+\beta+\gamma+4\eta<1$. Then T has a unique fixed point in H.

The purpose of this section is to extend Theorem 2.1 to the case of two mappings in a Hilbert space as the following.

Theorem 2.2. Let C be a closed subset of a Hilbert space H and S,T be mappings on C into itself satisfying:

$$\begin{aligned}\|Sx-Ty\|^2 \le\ & \alpha\ \|x-Sx\|^2+\beta\ \|y-Ty\|^2 \\ & +\gamma\ \|x-y\|^2+\delta \min\left\{\|x-Ty\|^2, \|y-Sx\|^2\right\} \\ & +\eta \frac{\|y-Sx\|^2}{1+\|x-Sx\|\|x-Ty\|},\end{aligned} \tag{2.2}$$

for all $x, y \in C$ where α, β, γ are non – negative reals with $\alpha+\beta+\gamma+4\eta<1$. Then S and T have a unique common fixed point in C.

Proof. Let $x_0 \in C$, we define a sequence $\{x_n\}$ as follows

$$x_{2n+1} = Sx_{2n} \ , x_{2n+2} = Tx_{2n+1} \ , n = 0, 1, 2, ...$$

From (2.2), we have

$$\begin{aligned}
\|x_{2n+1} - x_{2n}\|^2 &= \|Sx_{2n} - Tx_{2n-1}\|^2 \\
&\le \alpha \|x_{2n} - Sx_{2n}\|^2 + \beta \|x_{2n-1} - Tx_{2n-1}\|^2 \\
&\quad + \gamma \|x_{2n} - x_{2n-1}\|^2 \\
&\quad + \delta \min\left\{\|x_{2n} Tx_{2n-1}\|^2, \|x_{2n-1} - Sx_{2n}\|^2\right\} \\
&\quad + \eta \frac{\|x_{2n-1} - Sx_{2n}\|^2}{1 + \|x_{2n} - Sx_{2n}\| \|x_{2n} - Tx_{2n-1}\|} \\
&= \alpha \|x_{2n} - x_{2n+1}\|^2 + \beta \|x_{2n-1} - x_{2n}\|^2 + \gamma \|x_{2n} - x_{2n-1}\|^2 \\
&\quad + \delta \min\left\{\|x_{2n} - x_{2n}\|^2, \|x_{2n-1} - x_{2n+1}\|^2\right\} \\
&\quad + \eta \frac{\|x_{2n-1} - x_{2n+1}\|^2}{1 + \|x_{2n} - x_{2n+1}\| \|x_{2n} - x_{2n}\|} \\
&\le \alpha \|x_{2n} - x_{2n+1}\|^2 + \beta \|x_{2n-1} - x_{2n}\|^2 + \gamma \|x_{2n} - x_{2n-1}\|^2 \\
&\quad + \eta \|x_{2n-1} - x_{2n+1}\|^2
\end{aligned}$$

Thus we have

$$\|x_{2n} - x_{2n+1}\|^2 \le \frac{\beta + \gamma + 2\eta}{1 - \alpha - 2\eta} \|x_{2n-1} - x_{2n}\|^2.$$

Putting $k = \frac{\beta + \gamma + 2\eta}{1 - \alpha - 2\eta} < 1$

Then, we have

$$\|x_{2n} - x_{2n+1}\|^2 \le k \|x_{2n-1} - x_{2n}\|^2.$$

Proceeding in this way, we get

$$\|x_{2n} - x_{2n+1}\|^2 \le k^n \|x_0 - x_1\|^2 , \ n = 1, 2, ...$$

For any positive integer p, one gets :

$$\begin{aligned}\|x_n - x_{n+p}\| &\leq \|x_n - x_{n+1}\| + \|x_{n+1} - x_{n+2}\| \\ &\quad + ...+ \|x_{n+p-1} - x_{n+p}\| \\ &\leq (k^n + k^{n+1} + ...+ k^{n+p-1})\|x_0 - x_1\| \\ &\leq \frac{k^n}{1-k}\|x_0 - x_1\| \quad \to 0.\end{aligned}$$

Thus $\|x_n - x_{n+p}\| \to 0 \; as\, n \to \infty.$

Hence $\{x_n\}$ is a Cauchy sequence in C. Since C is closed subset of H, then there exists an element $v \in C$ such that

$$\lim_{n \to \infty} x_n = v.$$

Now further, we have

$$\begin{aligned}\| v - Tv\|^2 &= \|(v - x_{2n+1}) + (x_{2n+1} - Tv)\|^2 \\ &\leq \| v - x_{2n+1}\|^2 + \|x_{2n+1} - Tv\|^2 \\ &\quad + 2\operatorname{Re} < v - x_{2n+1}, x_{2n+1} - Tv > \\ &\leq \| v - x_{2n+1}\|^2 + \| Sx_{2n} - Tv\|^2 \\ &\quad + 2\operatorname{Re} < v - x_{2n+1}, x_{2n+1} - Tv > \\ &\leq \|v - x_{2n+1}\|^2 + \alpha\|x_{2n} - x_{2n+1}\|^2 \\ &\quad + \beta\|v - Tv\|^2 + \gamma\|x_{2n} - v\|^2 \\ &\quad + \delta \min\left\{\|x_{2n} - Tv\|^2, \|v - x_{2n+1}\|^2\right\} \\ &\quad + \eta\frac{\|v - x_{2n+1}\|^2}{1 + \|x_{2n} - x_{2n+1}\|\|x_{2n} - Tv\|} \\ &\quad + 2\operatorname{Re} < v - x_{2n+1}, x_{2n+1} - Tv >.\end{aligned}$$

As $n \to \infty, x_{2n} \to v, x_{2n+1} \to v,$

We have

$$2\,\mathrm{Re} < v - x_{2n+1}, x_{2n+1} - Tv > \to 0.$$

Then

$$\| v - Tv \|^2 \le \beta \| v - Tv \|^2.$$

This implies that v =T v, since $\beta < 1$.

Similarly we get v = S v. Then v is a common fixed point of S and T.

For the uniqueness, let $u \in C$ be another fixed point of S and T, where $u \neq v$, then

$$\|u - v\|^2 = \|Sv - Tu\|^2$$
$$= \alpha \| v - Sv\|^2 + \beta \| u - Tu\|^2 + \gamma \| u - v\|^2$$
$$+ \delta \min\left\{\|v - Tu\|^2, \|u - Sv\|^2\right\} + \eta \frac{\|u - Sv\|^2}{1 + \|v - Sv\|\|v - Tu\|}$$
$$\le (\gamma + \delta + \eta)\|u - v\|^2,$$

Since $\gamma + \delta + \eta < 1$, so u = v i.e., the common fixed point is unique.

Next, we extend Theorem 2.2 to the case of pair $S^p \text{ and } T^q$, where p and q are some positive integers and to the case of a sequence of mappings satisfying the same contractive condition (2.2).

Theorem 2.3. Let C be a closed subset of a Hilbert space H and S,T be mappings on C into itself satisfying

$$\|S^p x - T^q y\|^2 \le \alpha \|x - S^p x\|^2 + \beta \|y - T^q y\|^2$$
$$+ \gamma \|x - y\|^2 + \delta \min\left\{\|x - T^q y\|^2, \|y - S^p x\|^2\right\} \qquad (2.3)$$
$$+ \eta \frac{\|y - S^p x\|^2}{1 + \|x - S^p x\|\|x - T^q y\|},$$

for all $x, y \in C,\ p,\ q \in (0,1)$, where $\alpha, \beta, \gamma, \delta$ and η are nonnegative reals with $\alpha + \delta + \eta < 1$. Then S and T have a unique common fixed point.

Proof. Since $S^p and T^q$ satisfies all the conditions of theorem 2.2. Hence $S^p and T^q$ have a unique common fixed point, we assume that they have a common fixed point v

$$S^p v = v \Rightarrow S(S^p v) = Sv,$$

$$S^p(Sv) = Sv.$$

$$if \;\; Sv = x_0 \;, then \;\; S^p(x_0) = x_0,$$

So S v is a fixed point of S^p. Similarly we can show that T v is a fixed point of T^q i.e., $T^q(Tv) = Tv$. Now we have

$$\| v - Tv\|^2 = \left\|S^p v - T^q(Tv)\right\|^2$$

$$\leq \alpha\left\|v - S^p v\right\|^2 + \beta\left\|Tv - T^q(Tv)\right\|^2 + \gamma\|v - Tv\|^2$$

$$+\delta \min\left\{\left\|v - T^q(Tv)\right\|^2, \left\|Tv - S^p v\right\|^2\right\}$$

$$+\eta \frac{\left\|Tv - S^p v\right\|^2}{1 + \left\|v - S^p v\right\| \left\|v - T^q(Tv)\right\|}$$

$$\leq \gamma\|v - Tv\|^2 + \delta \min\left\{\|v - Tv\|^2, \|Tv - v\|^2\right\}$$

$$+\eta\|Tv - v\|^2.$$

Then we have:

$$\| v - Tv\|^2 \leq (\gamma + \delta + \eta)\|Tv - v\|^2 \;\;, \gamma + \delta + \eta < 1,$$

So, we have v = Tv.

On the same way we can prove that S v =v. So v is a common fixed point of S and T.

To prove the uniqueness, let $v \neq w$ be another common fixed point of S and T. Then clearly w is also a common fixed point of $S^p and T^q$. So from Theorem 2.3 $S^p and T^q$ have a common fixed point. Therefore w= v. Hence S and T have a unique common fixed point.

Hence we have proved that if x_0 is unique common fixed point of S^p and T^q for all p, q > 0, then x_0 is unique common fixed point of S and T.

Theorem 2.4. Let C be a closed subset of a Hilbert space H and let $\{F_i\}$ be a sequence of mapping on C converging point wise to F satisfying

$$\|F_ix-F_iy\|^2 \le \alpha\|x-F_ix\|^2+\beta\|y-F_iy\|^2+\gamma\|x-y|$$
$$+\delta\min\left\{\|x-F_iy\|^2,\|y-F_iy\|^2\right\}$$
$$+\eta\frac{\|y-F_ix\|^2}{1+\|x-F_ix\|\;\|x-F_iy\|},$$

for all x, y in C, where $\alpha,\beta,\gamma,\delta$ and η are nonnegative reals in [0,1] with $\alpha+\beta+\gamma+4\eta<1$. If F_i has a fixed point v_i and F has a fixed point v. Then the sequence $\{v_n\}$ converges to v.

Proof. Since it is given $F_iv_i=v_i$ and $Fv=v$.

Now

$$\|v-v_n\|^2=\|Fv-F_nv_n\|^2$$
$$=\|(Fv-F_nv)+(F_nv-F_nv_n)\|^2$$
$$\le\|Fv-F_nv\|^2+\|F_nv-F_nv_n\|^2$$
$$+2\operatorname{Re}<Fv-F_nv,F_nv-F_nv_n>$$
$$\le\|Fv-F_nv\|^2+\alpha\|v-F_n\|^2+\beta\|v_n-F_nv_n\|^2$$
$$+\gamma\|v-v_n\|^2+\delta\min\left\{\|v-F_nv\|^2,\|v_nF_nv\|^2\right\}$$
$$+\eta\frac{\|v_n-F_nv\|^2}{1+\|v-F_nv\|\|v-F_nv_n\|}$$
$$+\operatorname{Re}<Fv-F_nv,F_nv-F_nv_n>.$$

Taking $n\to\infty$ and $F_nv\to Fv$, we have

$$\|v-v_n\|^2\le\gamma\|v-v_n\|^2+\eta\|v-v_n\|^2$$
$$\le(\gamma+\eta)\|v-v_n\|^2,$$
$$\gamma+\eta<1.$$

So $v_n \to v$ as $n \to \infty$.

COINCIDENCES FIXED POINT THEOREMS

Our main results in this section are the following theorems

Theorem 3.1. Let H be a Hilbert space and let $T, f : H \to H$ be such that :

(i) $T(H) \subset f(H)$

$$(ii)\ \| Tx - Ty\|^2 \le \alpha \max \left\{ \begin{array}{l} \|fx - fx\|^2, \|fx - Tx\|^2, \\ \|fy - Ty\|^2, \\ \dfrac{\|fx - Ty\|^2 + \|fy - Tx\|^2}{2} \end{array} \right\}, \tag{3.1}$$

for all x , y $\in$ H where $0 < \alpha < \frac{1}{8}$

(iii) f(H) or T(H) is a complete subspace of

(iv) T is asymptotically f - regular of some point x_0 in H.

Then T and f have a point of coincidence

Proof. Let $\{fx_n\}$ be an asymptotically T – regular sequence in H. Then by Parallelogram law we have

$$\begin{aligned} \|fx_n - fx_m\|^2 &\le 2\|fx_n - Tx_n\|^2 + 2\|Tx_n - fx_m\|^2 \\ &\le 2\|fx_n - Tx_n\|^2 + 4\|Tx_n - Tx_m\|^2 \\ &\quad + 4\|Tx_m - fx_m\|^2 \end{aligned} \tag{3.2}$$

Using (3.1) we have

$$\|fx_n - fx_m\|^2 \le 2\,\|fx_n - Tx_n\|^2 + 2\,\|Tx_n - fx_m\|^2$$

$$\le 2\,\|fx_n - Tx_n\|^2 + 4\,\|Tx_m - fx_m\|$$

$$+ 4\,\alpha \max\left\{\begin{array}{l}\|fx_n - fx_m\|^2, \|fx_n - Tx_n\|^2, \\ \|fx_m - Tx_m\|^2, \\ \dfrac{\|fx_n - Tx_m\|^2 + \|fx_m - Tx_n\|^2}{2}\end{array}\right\} \tag{3.3}$$

$$\le 2\|fx_n - fx_{n+1}\|^2 + 4\,\|fx_{m+1} - fx_m\|$$

$$+ 4\,\alpha \max\left\{\begin{array}{l}\|fx_n - fx_m\|^2, \|fx_n - fx_{n+1}\|^2, \|fx_m - fx_{m+1}\|^2, \\ \dfrac{2\|fx_n - fx_m\|^2 + 2\|fx_m - fx_{m+1}\|^2}{2} + \\ \dfrac{2\|fx_m - fx_n\|^2 + 2\|fx_n - fx_{n+1}\|^2}{2}\end{array}\right\}.$$

Taking limit as $n,m \to \infty$, and using asymptotically T – regular of $\{fx_n\}$ gives

$$\|fx_n - fx_m\|^2 \le 4\,\alpha \max\left\{\|fx_n - fx_m\|^2, 0, 0, 2\|fx_n - fx_m\|^2\right\}$$

$$\le 8\,\alpha\|fx_n - fx_m\|^2, \quad 0 < \alpha < \frac{1}{8}$$

Hence we have

$$\lim_{n,m\to\infty} \|fx_n - fx_m\| = 0.$$

It follows that $\{fx_n\}$ is a Cauchy sequence in H. If f(H) is a complete subspace of H, there exists a point $p \in H$ such that:

$fx_n \to p = fu$ (this also holds if T(H) is complete with $p \in T(H)$).

We claim that u is a coincidence point of f and T. If not $\|u - Tu\| > 0$. From (ii), we obtain

$$\|fu - Tu\|^2 = \|p - fx_{n+1} + fx_{n-1} - Tu\|^2$$
$$\leq 2\|p - fx_{n+1}\|^2 + 2\|fx_{n+1} - Tu\|^2$$
$$\leq 2\|p - fx_{n+1}\|^2 + 2\|Tx_n - Tu\|^2$$
$$\leq 2\|p - fx_{n+1}\|^2$$
$$+ 2\alpha \max\left\{ \begin{array}{l} \|fx_n - fu\|^2, \|fx_n - Tx_n\|^2, \|fu - Tu\|^2, \\ \dfrac{\|fx_n - Tx_n\|^2 + \|fu - Tu\|^2}{2} \end{array} \right\}$$

As $n \to \infty$ we get

$$\| fu - Tu\|^2 = 2\alpha \max\{0, 0, \| fu - Tu \|^2, \frac{1}{2}\|fu - Tu\|^2\}.$$

Hence

$$\|fu - Tu\|^2 \leq 2\alpha\|fu - Tu\|^2$$

a contradiction and so p = f u = T u is a point of coincidence of f and T.

Theorem 3.2. Let H be a Hilbert space and let $T, f : H \to H$ be such that:

(a) $T(H) \subset f(H)$

$$(b)\ \|Tx - Ty\|^2 \leq \alpha \max\left\{ \begin{array}{l} \|fx - fy\|^2, \|fx - Tx\|^2, \\ \|fy - Ty\|^2, \\ \dfrac{\|fx - Ty\|^2 + \|fy - Tx\|^2}{2} \end{array} \right\}, \qquad (3.4)$$

for all $x, y \in H$ where $0 < \alpha < 1$.

Then f, T have at least a unique point of coincidence.

Proof. Assume there exist p, p^* in H such that

P = f u = T u and $p^* = fu^* = Tu^*$ for some $u, u^* \in H$

From (3.4) we obtain

$$\left\|p-p^*\right\|^2 = \left\|Tu-Tu^*\right\|$$
$$\leq \alpha \max\left\{ \begin{array}{l} \left\|fu-fu^*\right\|^2, \left\|fu-Tu\right\|^2, \left\|fu^*-Tu^*\right\|^2, \\ \dfrac{\left\|fu-Tu^*\right\|^2+\left\|fu^*-Tu\right\|^2}{2} \end{array} \right\}$$
$$\leq \alpha \max\left\{ \left\|p-p^*\right\|^2, 0, 0, \frac{\left\|p-p^*\right\|^2+\left\|p^*-p\right\|^2}{2} \right\}$$
$$\leq \alpha \max\left\{ \left\|p-p^*\right\|^2, 0, 0, \left\|p-p^*\right\|^2 \right\} \leq \alpha \left\|p-p^*\right\|^2,$$

we deduce that $p^* = p$.

Remark. 3.1. If we put $f = Id$ *(identity map)* in Theorem 3.1and in Theorem 3.2, we obtain Theorems 3. 1 and Theorem 3.2 in [1].

Let $F_i : [0,\infty) \to [0,\infty)$ be functions such that $F_i(0)$ and F_i is continuous at 0 (i=1,2). Ciric [5] studied necessary conditions to obtain a fixed point result of asymptotically regular mappings on complete metric spaces. M. Abbas and H.Aydi [1] extended the results of Ćirić [5] to the case of two mappings satisfying a generalized contractive conditions in a metric space and they proved the following theorem.

Theorem 3.3. [5]. Let (X, d) be a metric space. Let $T, f: X \to X$ be such that $T(X) \subset f(X)$. Assume that the following condition holds:

$$\begin{aligned} d(Tx,Ty) &\leq a_1 F_1[\min\{d(fx,Tx), f(fy,Ty)\}] \\ &+ a_2 F_2[d(fx,Tx)d(fy,Ty)] + a_3 d(fx,fy) \\ &+ a_4[d(fx,Tx)+d(fy,Ty)] + a_5[d(fx,Ty)+d(fy,Tx)], \end{aligned} \tag{3.5}$$

for all $x, y \in X$,where $i = 1,2,......5, a_i \geq 0$ such that for arbitrary fixed k > 0, $0 < \lambda_1 < 1$ and $0 < \lambda_2 < 1$ and $a_1 a_2 \leq k$. If $f(X)$ or $T(X)$ is a complete subspace of X and T is asymptotically f- regular at some point x_0 in X, then T and f have a unique point of coincidence.

In the following theorem we prove a similar result using a contractive condition (3.5) in a Hilbert space.

Theorem 3.4. Let H be a Hilbert space. Let $f, T: H \to H$ be such that:

(1) $T(H) \subset f(H)$

$$(2)\ \|Tx-Ty\|^2 \le a_1F_1[\min\{\|fx-Tx\|^2, \|fy-Ty\|^2\}]+$$
$$a_2F_2[\|fx-Tx\|\|fy-Ty\|]+a_3\|fx-fy\|^2+a_4[\|fx-Tx\|^2 \tag{3.6}$$
$$+\|fy-Ty\|^2]+a_5[\|fx-Ty\|^2+\|fy-Tx\|^2],$$

for all x, $y \in H$ where $a_i \ge o, i=1,2,....,5$ such that for arbitrary fixed $k>0, 0<\lambda_1<\frac{1}{2}$ and $0<\lambda_2<\frac{1}{8}$, we have $a_4+a_5 \le \lambda_1$ $a_3+a_5<\lambda_2$ and $a_1a_2<k$. If $f(H) \subset T(H)$ is a complete subspace of H and if T is asymptotically f –regular at some point x_0 in X. Then f, T have a point of coincidence.

Proof. Let x_0 be an arbitrary point in H and let $\{fx_n\}$ be a T-sequence with initial point x_0. Since T is asymptotically f-regular mappings at x_0 therefore $\|fx_n - fx_{n+1}\| \to 0$ as $n \to \infty$. Now for m > n, we have

$$\|fx_n - fx_m\|^2 = \|(fx_n - Tx_n) + (Tx_n - fx_m)\|^2$$
$$\le 2\|fx_n - Tx_n\|^2 + 2\|Tx_n - fx_m\|^2$$
$$\le 2\|fx_n - Tx_n\|^2 + 4\|Tx_n - Tx_m\|^2 + 4\|Tx_m - fx_m\|^2$$
$$\le 2\|fx_n - fx_{n+1}\|^2 + 4\|fx_m - fx_{m+1}\|^2$$
$$+4a_1F_1[\min\{\|fx_n - fx_{n+1}\|^2, \|fx_m - fx_{m+1}\|^2\}]$$
$$+4a_2F_2[\|fx_n - fx_{n+1}\|^2\|fx_m - fx_{m+1}\|^2]$$
$$+4a_3\|fx_n - fx_m\|^2 + +4a_4[\|fx_n - fx_{n+1}\|^2 + \|fx_m - fx_{m+1}\|^2]$$
$$+4a_5[\|fx_n - fx_m\|^2 + 2\|fx_m - fx_{m+1}\|^2 + 2\|fx_n - fx_{n+1}\|^2].$$

Thus we obtain that

$$1-4(a_3+a_5)\ \|fx_n - fx_m\|^2$$
$$\le 2\|fx_n - fx_{n+1}\|^2 + 4\|fx_m - fx_{m+1}\|^2 + 4\|Tx_m - fx_m\|^2$$
$$+4a_1F_1[\min\{\|fx_n - fx_{n+1}\|^2, \|fx_m - fx_{m+1}\|^2\}] \tag{3.7}$$
$$+4a_2F_2[\|fx_n - fx_{n+1}\|^2\|fx_m - fx_{m+1}\|^2]$$
$$+4a_3\|fx_n - fx_m\|^2$$
$$+8a_5[\|fx_m - fx_{m+1}\|^2 + \|fx_n - fx_{n+1}\|^2].$$

Since T is asymptotically f–regular and F_1, F_2 are continuous at zero, then the right hand of the inequality (3.7) tends to zero as $m, n \to \infty$. Thus

$$\lim_{n,m\to\infty} \|fx_n - fx_m\| = 0.$$

It follows that $\{fx_n\}$ is a Cauchy sequence in H. If f(H) is a complete subsequence of H there exists u, p in H such that $fx_n \to p = fu$ (this holds also if T(H) is complete with $p \in T(H)$).

We claim that u is a coincidence point of f and T. If not $\|u - Tu\| > 0$. From (3.6) we obtain:

$$\begin{aligned}
&\|fu - Tu\|^2 = \|p - Tu\|^2 \le 2\|p - fx_{n+1}\|^2 + 2\|fx_{n+1} - Tu\|^2 \\
&\le 2\|p - fx_{n+1}\|^2 + 2\|Tx_n - Tu\|^2 \\
&\le 2\|p - fx_{n+1}\|^2 + 2a_1 F_1[\min\left\{\|fx_n - Tx_n\|^2, \|fu - Tu\|^2\right\} \\
&+2a_2F_2[\|fx_n - Tx_n\|\|fu - Tu\|\ ] + 2a_3\|fx_n - fu\|^2 \\
&+2a_4[\|fx_n - Tx_n\|^2 + \|fu - Tu\|^2] \\
&+2a_5[\|fx_n - Tu\|^2 + \|fu - Tx_n\|^2],
\end{aligned}$$

which yields that

$$\begin{aligned}
\|fu - Tu\|^2 \le\ & 2\|p - fx_{n+1}\|^2 \\
&+ 2a_1 F_1[\min\left\{\|fx_n - fx_{n+1}\|^2, \|fu - Tu\|^2\right\}] \\
&+2a_2F_2[\|fx_n - fx_{n+1}\|\|fu - Tu\|\ ] + 2a_3\|fx_n - fu\|^2 \\
&+ 2a_4[\|fx_n - fx_{n+1}\|^2 + \|fu - Tu\|^2] \\
&+ 2a_5[\|fx_n - Tu\|^2 + \|fu - fx_{n+1}\|^2].
\end{aligned}$$

Taking limit as $n \to \infty$, we get

$$\|fu - Tu\|^2 \le 2(a_4 + a_5)\|fu - Tu\|^2 \le 2\lambda_1\|fu - Tu\|^2, 0 < \lambda_1 < \frac{1}{2},$$

a contradiction and so p = fu = T u is a point of a coincidence of f and T.

Lemma 3.5. Let H be a Hilbert space. Let $f,T:H\to H$ be such that:

(1) $T(H)\subset f(H)$

(2) $\|Tx-Ty\|^2 \le a_1F_1[\min\{\|fx-Tx\|^2,\|fy-Ty\|^2\}]$

$+a_2F_2[\|fx-Tx\|\|fy-Ty\|]+a_3\|fx-fy\|^2+a_4[\|fx-Tx\|^2$

$+\|fy-Ty\|^2]+a_5[\|fx-Ty\|^2+\|fy-Tx\|^2]$,

for all $x,y\in H$, where $a_i\ge 0, i=1,2,....,5$ such that for arbitrary fixed $k>0, a_1\, a_2 \le k$ and $a_3+2a_5<1$. Then f, T have at most a unique coincidence point

Proof. Assume that there exists, p,p^* in H such that $p=fu=Tu$ and $p^*=fu^*=Tu^*$ for some $u, u^*\in H$ such that

$$\|p-p^*\|^2=\|Tu-Tu^*\|^2$$

$$\le a_1F_1[\min\{\|fu-Tu\|^2,\|fu^*-Tu^*\|^2\}$$

$$+a_2F_2[\|fu-Tu\|\|fu^*-Tu^*\|]+a_3\|fu-fu^*\|^2$$

$$+a_4[\|fu-Tu\|^2+\|fu^*-Tu^*\|^2]+a_5[\|fu-Tu^*\|^2+\|fu^*-Tu\|^2].$$

Hence we obtain

$$\|p-p^*\|^2\le(a_3+2a_5)\|p-p^*\|^2,\quad a_3+2a_5<1.$$

This shows that $p=p^*$.

From Theorem 3.4 and Lemma 3.5 we obtain the following theorem

Theorem 3.6. Let H be a Hilbert space and let f, T be mappings on H into H such that $T(H)\subset f(H)$. Assume that T and f satisfy condition 3.6 for all $x,y\in H$. If f(H) or T(H) is a complete subspace of H such that (T, f) is weakly compatible, then T and f have a unique common fixed point provided that T is asymptotically f- regular at some point x_0 in H.

As a consequence of Theorem 3.4, Lemma 3.5 and Theorem 3.6 we obtain the following corollary.

Corollary 3.7. Let H be a Hilbert space. Let $T: H \to H$ be such that the following condition holds:

$$\begin{aligned}\|Tx-Ty\|^2 \le a_1F_1[\min\left\{\|x-Tx\|^2, \|y-Ty\|^2\right\}] \\ +a_2F_2[\|x-Tx\|\|y-Ty\|]+a_3\|x-fy\|^2 \\ +a_4[\|x-Tx\|^2+\|y-Ty\|^2] \\ +a_5[\|x-Ty\|^2+\|y-Tx\|^2],\end{aligned} \tag{3.8}$$

for all $x,y \in H$, where $a_i \ge 0 (i=1,2,...5)$ such that for an arbitrary fixed $k>0$, $0<\lambda_1<\frac{1}{2}, 0<\lambda_2<\frac{1}{4}$ we have $a_4+a_5<\lambda_1, a_3+a_5<\lambda_2, a_1a_2 \le k$. If T is asymptotically regular at some point x_0 in H. Then T has a unique fixed point.

Taking $a_1 = a_2 = 0$ in the inequality 3.6, we have the following corollary.

Colorary 3.8. Let H be a Hilbert space. Let $f,T:H \to H$ be such that $T(H) \subset f(H)$.Assume that the following condition holds:

$$\begin{aligned}\|Tx-Ty\|^2 \le a_3\|fx-fy\|^2+a_4[\|fx-Tx\|^2+\|fy-Ty\|^2] \\ +a_5[\|fx-Ty\|^2+\|fy-Tx\|^2],\end{aligned}$$

for all $x,y \in H$, where $a_i \ge 0 (i=3,...5)$ such that for arbitrary fixed $k>0$, $0<\lambda_1<\frac{1}{2}, 0<\lambda_2<\frac{1}{4}$, we have $a_4+a_5<\lambda_1, a_3+a_5<\lambda_2$, $a_1a_2 \le k$. If f(H) or T(H) is a complete subspace of H and if T is asymptotically regular at some point x_0 in H. Then T and f have a point. of coincidence.

We give an example to support our results

Example 3.9. Let X= $[0,\infty)$ and let Le t $T: H \to H$ be defined as

$$Tx = \frac{x}{4}, \quad fx = x.$$

Let x_0=1 and the sequence $\{x_n\}_{n=1}^{n=\infty}$ be given by $x_n = \frac{1}{2^n}$. Note that $\{fx_n\}$ is a T-sequence with initial point x_0. Since $\|fx_n - fx_{n+1}\| \to 0$ as $n \to \infty$, the mapping T is asymptotically f – regular at the point x_0. Also T(X) $\subset f(X)$, T(X) is a complete and the inequality (3.6) holds for all $x,y \in X$ with

$$F_1 = F_2 = 1 \ , a_1 = a_2 = a_4 = a_5 = 0 \ , a_3 = \frac{1}{16}.$$

Thus $f, T: X \to X$ satisfy all conditions of Theorem 3.4. Moreover u=0 is the common fixed point of f and T.

Let $CF(T, f)$ denote the set of all common fixed points of f and T. Now we have the following result on the continuity on the set of common fixed points

Theorem 3.10. Let H be a Hilbert space. Assume that $f, T: H \to H$ satisfy condition (3.6) for all $x, y \in H$. If $CF(T, f) \neq \phi$. Then T is continuous at $p \in CF(T, f)$ whenever f is continuous at p.

Proof. Fix $p \in CF(T, f)$.Let $\{z_n\}$ be any sequence in H converging to p. Then by taking $y := z_n$ and z:=p in (3.6) we get

$$\begin{aligned}\|Tx - Tz_n\|^2 \le a_1 F_1[\min\left\{\|fp - Tp\|^2, \|fz_n - Tz_n\|^2\right\}] \\ + a_2 F_2[\|fp - Tp\|\|fz_n - Tz_n\|] + a_3\|fp - fz_n\|^2 \\ + a_4[\|fp - Tp\|^2 + \|fz_n - Tz_n\|^2] \\ + a_5[\|fp - Tz_n\|^2 + |fz_n - Tp\|^2],\end{aligned}$$

which in view of T p = f p, we obtain

$$\begin{aligned}\|Tp - Tz_n\|^2 \le a_3\|fp - fz_n\|^2 + a_4\|Tp - Tz_n\|^2 \\ + a_5[\|Tp - Tz_n\|^2 + \|fz_n - Tp\|^2].\end{aligned}$$

Now by letting $n \to \infty$, we get

$$\overline{\lim_{n\to\infty}} \| \ Tp - Tz_n \ \|^2 \le (a_4 + a_5) \overline{\lim_{n\to\infty}} \| \ Tp - Tz_n \ \|^2,$$

whenever f is continuous at p. The last inequality is true only if

$$\overline{\lim_{n\to\infty}} \|Tp - Tz_n\|^2 = 0.$$

We get that $Tz_n \to Tp \ as \ n \to \infty$.

WELL – POSEDNESS

The notion of well - posedness of a fixed point problem has generated much interest to several mathematicians, for example [2, 3, 8, 11, 12, 13]. Here, we study well – posedness of a common fixed point problem.

Definition 4.1. Let H be a Hilbert space and $f: H \to H$ be a mapping. The fixed point problem of f is said to be well posed if

(i) f has a unique fixed point x_0 in H
(ii) for any sequence $\{x_n\} \in H$, $\lim_{n\to\infty} \|x_n - fx_n\| = 0$ we have $\lim_{n\to\infty} \|x_n - x_0\| = 0$.

Definition 4.2. A common fixed point problem of self-maps f and T on H, CFP(f,T,H) is called well- posed if CF(f, T) (the set of all common fixed points of f and T) is singletion and for any sequence $\{x_n\}$ in H with $x^* \in CF(f,T)$ and $\lim_{n\to\infty} \| fx_n - x_n \| = \lim_{n\to\infty} \| Tx_n - x_n \| = 0$ implies $x^* = \lim_{n\to\infty} x_n$.

Theorem 4.1. Suppose that T and f be self – maps on H as in Theorem 3.4 and Lemma 3.1. Then the common fixed problem of f and T is well posed.

Proof. From Theorem 3.4 and Lemma 3.1, the mappings f and T have a unique common fixed point, say u $\in H$. Let $\{x_n\}$ be a sequence in H and $\lim_{n\to\infty} \| fx_n - x_n \| = \lim_{n\to\infty} \| Tx_n - x_n \| = 0$. With loss of generality, we may suppose that $u \neq x_n$ for every nonnegative integer n. Then having in mind fu =T u and from triangle inequality (3.6), we have,

$$\|u - x_n\|^2 = \|Tu - x_n\|^2 \le 2\|Tu - Tx_n\|^2 + 2\|Tx_n - x_n\|^2$$
$$\le 2\|Tx_n - x_n\|^2 + a_1 F_1[\min\{\|fx_n - Tx_n\|^2, \|fu - Tu\|^2\}]$$
$$+ a_2 F_2[\|fx_n - Tx_n\| \|fu - Tu\|] + a_3 \|fx_n - fu\|^2$$
$$+ a_4[\|fx_n - Tx_n\|^2 + \|fu - Tu\|^2] + a_5[\|fx_n - Tu\|^2 + \|u - Tx_n\|^2]$$
$$\le 2\|Tx_n - x_n\|^2 + a_3[2\|fx_n - x_n\|^2 + 2\|x_n - u\|^2]$$
$$+ a_4[2\|fx_n - x_n\|^2 + 2\|x_n - Tx_n\|^2] + a_5[2\|fx_n - x_n\|^2$$
$$+ 4\|x_n - u\|^2 + 2\|x_n - Tx_n\|^2].$$

Letting $n \to \infty$, we get that $\overline{\lim_{n\to\infty}} \| u - x_n \| = 0$. We deduce, $x_n \to u$ as $n \to \infty$. This completes the proof of Theorem.

ACKNOWLEDGEMENT

The author would like to express his thanks to the referees and editors for helpful comments and suggestions.

REFERENCES

1. Abbas, M. and Aydi, H. "On common fixed point of generalized contractive mappings in metric spaces", surveys in Mathematics and its Applications, Vol.7, 39-47, (2012).
2. Akkouchi, M., Well posedness of the fixed point problem for certain asymptotically regular mappings", AnnalasMathematicae Silesianae, 23, 43-52, (2009).
3. Akkouchi, M. and Popa, V., "Well posedness of the fixed point problem for mappings satisfying an implicit relations", Demonstratio Mathematica, Vol.23, No.4, 923-929, (2010).
4. Bhopal, S. S., "Some fixed point theorems in Hilbert spaces for self mappings", American Journal of Mathematical Modelling Vol.2 No.3, 17-25, (2012).
5. Ćirić, Lj.B., "fixed points of asymptotically regular mappings", Mathematical Communations,
6. Ćirić, Lj. B., "Generalized contractions and fixed point theorems", Publ. Inst. Math. (Beograd), 12(26), 19-26, (1971).
7. Guay, M. D., and Singh, K.L., "fixed points of asymptotically regular mappings", Math. Vesnik, 35, 101-106 (1983).
8. Lahiri, B. K., "Well-Posedness and certain classes of operators", Demonstratio Math., 38, 170-176 (2005).
9. Patel, S. T., Garg, S. and Bhardwaj, R. , Some results concerning fixed point in Hilbert space", Journal ofEngineering Research and Applications (IJERA), Vol.2 ssue.4, 1459-1461, (2012).
10. Patel, M.K. and Deheri, G.M., "Some fixed point theorems on Hilbert spaces", GIT- Journal of Engineering andTechnology Vol.6, 2013.
11. Popa, V., "Well- Posedness of Fixed problem in compact Metric space", Bull. Univ. Petrol-Gaze, Ploicsti, sec.Mat Inform. Fiz. 60(1), 1-4, (2008).
12. Popa, V., "Well- Posedness of fixed point problem in orbitally complete metric paces",Stud.Cerc.St.Ser.Mat.Univ.16 (2006), supplement, Proceedings of ICMI 45, Bacau, Sept. 18-20, 209-214 (2006).
13. Reich, S., and Zaslawski, A.T., Well- Posedness of fixed point problems, Far East J. Math. sci, Special volume, part III, 393-401 (2011).
14. Sharma, P.L., and Yuel, A.K., fixed point theorems under Asymptotic regularity at a point", Math. Sem. Notes35, 181-190, (1982).

19

ON THE ENESTRÖM-KAKEYA THEOREM

Gulshan Singh*

* Department of Mathematics, Jammu and Kashmir State Government of Education, India

ABSTRACT

In this paper, we prove some extensions of the Eneström-Kakeya theorem by relaxing the hypothesis in different ways which in turn generalizes a result of Aziz and Zargar [Some extensions of Eneström-Kakeya Theorem, Glasnik Matematički, 31(1996), 239-244].

INTRODUCTION AND STATEMENT OF RESULTS

Finding the roots of a polynomial is a long standing classical problem [3,7]. The various results in the analytic theory of polynomials concerning the number of zeros in a region have been frequently investigated. It is an interesting area of research for engineers as well as mathematicians and many results on the same topic are available in the literature. Over last five decades, a large number of research papers, e.g, [1,4,5,6,8,9,13,14] and monographs [10,12,15] have been published. Polynomials in various forms have recently come under extensive revision because of their applications in linear control systems, signal processing, electrical networks, coding theory and several areas of physical sciences, where among others location of zeros and stability problems arise in a natural way. Existing results in the literature also show that there is need to find bounds for special polynomials, for example, those having restrictions on the coefficient and there is always need for refinement of results in this subject.

The following result is well known in the theory of the distribution of zeros of polynomials.

Theorem A. *If* $P(z) := \sum_{j=0}^{n} a_j z^j$ *is a polynomial of degree n such that*

$$a_n \geq a_{n-1} \geq a_{n-2} \geq \ldots \geq a_1 \geq a_0 > 0,$$

Then $P(z)$ does not vanish in $|z| > 1$.

If we apply this result to the polynomial $P(kz), k > 0$, then it can be restated as:

Theorem B. *Let* $P(z) := \sum_{j=0}^{n} a_j z^j$ *be a polynomial of degree n such that*

$$a_n k^n \geq a_{n-1} k^{n-1} \geq a_{n-2} k^{n-2} \geq \ldots \geq a_1 k \geq a_0 > 0,$$

then all the zeros of $P(z)$ lie in

$$|z| \leq k. \tag{1}$$

The Eneström-Kakeya theorem is a very strong tool to find the region in the complex plane containing all the zeros of a class of polynomials. It has been used to analyze overflow oscillation of discrete-time dynamical system [11], to investigate the properties of orthogonal wavelets [9], to determine the asymptotic behavior of zeros of the Daubechies filter [16], in addition for application to a model of high energy collisions [2].

In the literature, [1,4,5,6,14], there exist extensions and generalizations of Eneström- Kakeya theorem.

Joyal et al. [8] obtained the following generalization, by considering the coefficients to be real, instead of being only positive.

Theorem C. *Let* $P(z) := \sum_{j=0}^{n} a_j z^j$ *be a polynomial of degree n such that*

$$a_n \geq a_{n-1} \geq a_{n-2} \geq \ldots \geq a_1 \geq a_0,$$

then all the zeros of $P(z)$ lie in

$$|z| \leq \frac{1}{|a_n|}\{a_n - a_0 + |a_0|\}.$$

Aziz and Zargar [1] also relaxed the hypothesis of Eneström-Kakeya theorem in a different way and proved the following results.

Theorem D. *Let* $P(z) := \sum_{j=0}^{n} a_j z^j$ *be a polynomial of degree n such that for some* $k \geq 1$,

$$ka_n \geq a_{n-1} \geq a_{n-2} \geq \ldots \geq a_1 \geq a_0 > 0,$$

then all the zeros of $P(z)$ lie in

$$|z + k - 1| \leq k.$$

Theorem E. Let $P(z) := \sum_{j=0}^{n} a_j z^j$ be a polynomial of degree n such that for some $k \geq 1$,

$$ka_n \geq a_{n-1} \geq a_{n-2} \geq \ldots \geq a_1 \geq a_0,$$

then all the zeros of $P(z)$ *lie in*

$$|z + k - 1| \leq \frac{ka_n - a_0 + |a_0|}{|a_n|}.$$

In this paper, we prove more general result by relaxing the hypothesis in different ways which includes Theorem E as a special case.

Theorem 1. *Let* $P(z) := \sum_{j=0}^{n} a_j z^j$ *be a polynomial of degree n such that* $\alpha_j = a_j + ib_j$ *where* a_j *and* b_j, $j = 0,1,\ldots,n$ *are real numbers. If for some positive integers* $\lambda, \mu \leq n$, *and for some real numbers* $0 < \rho_1 \leq 1, 0 < \rho_2 \leq 1, k \geq 1$,

$$\begin{aligned} k^{n-\lambda+1} a_n &\geq k^{n-\lambda} a_{n-1} \geq k^{n-\lambda-1} a_{n-2} \ldots \geq k^2 a_{\lambda+1} \\ &\geq ka_\lambda \geq a_{\lambda-1} \geq \ldots a_1 \geq \rho_1 a_0, \\ k^{n-\mu+1} a_n &\geq k^{n-\mu} a_{n-1} \geq k^{n-\mu-1} a_{n-2} \ldots \geq k^2 a_{\mu+1} \\ &\geq ka_\mu \geq a_{\mu-1} \geq \ldots a_1 \geq \rho_2 b_0, \end{aligned}$$

then all the zeros of $P(z)$ lie in

$$|z+k-1| \le \frac{1}{|a_n|}\Bigg\{(a_n+b_n)-(\rho_1 a_0+\rho_2 b_0)+(1-\rho_1)|a_0| \\ +(1-\rho_2)|b_0|+(|a_0|+|b_0|) \\ +(k-1)\left(\begin{array}{l}\sum_{j=\lambda}^{n}\left(a_j+|a_j|\right) \\ +\sum_{j=\mu}^{n}\left(b_j+|b_j|\right)-|a_n|-|b_n|\end{array}\right)\Bigg\}. \tag{2}$$

Remark 1. Theorem E is a special case of Theorem 1, if we take all coefficients of $P(z)$ are real, $\rho_1 = 1$ and $\lambda = n$.

The following Corollary immediately follows from Theorem 1.

Corollary 1. *Let $P(z) := \sum_{j=0}^{n} \alpha_j z^j$ be a polynomial of degree n such that $\alpha_j = a_j + ib_j$ where a_j and b_j, $j = 0,1,\ldots,n$ are real numbers. If for some positive integers $\lambda, \mu \le n$, and for some real numbers $0 < \rho_1 \le 1, 0 < \rho_2 \le 1, k \ge 1$,*

$$a_n \ge a_{n-1} \ge a_{n-2} \ldots \ge a_{\lambda+1} \ge ka_\lambda \ge a_{\lambda-1} \ge \ldots \ge a_1 \ge \rho_1 a_0,$$

$$b_n \ge b_{n-1} \ge b_{n-2} \ldots \ge b_{\lambda+1} \ge kb_\mu \ge b_{\mu-1} \ge \ldots \ge a_1 \ge \rho_2 b_0,$$

then all the zeros of $P(z)$ lie in

$$|z+k-1| \le \frac{1}{|a_n|}\Bigg\{(a_n+b_n)-(\rho_1 a_0+\rho_2 b_0)+(1-\rho_1)|a_0| \\ +(1-\rho_2)|b_0|+(|a_0|+|b_0|) \\ +(k-1)\left(\begin{array}{l}\sum_{j=\lambda}^{n}\left(a_j+|a_j|\right) \\ +\sum_{j=\mu}^{n}\left(b_j+|b_j|\right)-|a_n|-|b_n|\end{array}\right)\Bigg\}.$$

If we choose $\lambda = \mu = n-1$ and $a_0, b_0 > 0$ in Theorem 1, we have the following:

Corollary 2. *Let* $P(z) := \sum_{j=0}^{n} \alpha_j z^j$ *be a polynomial of degree n such that* $\alpha_j = a_j + ib_j$ *where* a_j *and* b_j, $j = 0,1,\ldots,n$ *are real numbers. If for some real numbers* $0 < \rho_1 \le 1, 0 < \rho_2 \le 1, k \ge 1,$

$$k^2 a_n \ge k a_{n-1} \ge a_{n-2} \ge a_{n-3} \ge \ldots \ge a_1 \ge \rho_1 a_0 > 0,$$

$$k^2 b_n \ge k b_{n-1} \ge b_{n-2} \ge b_{n-3} \ge \ldots \ge b_1 \ge \rho_2 b_0 > 0,$$

then all the zeros of $P(z)$ *lie in*

$$|z + k - 1| \le \frac{1}{|a_n|} \left\{ \begin{matrix} 2((1-\rho_1)a_0 + (1-\rho_2)b_0) \\ +k(a_n + b_n) + 2(k-1)(a_{n-1} + b_{n-1}) \end{matrix} \right\}.$$

Also, if we put $\lambda = \mu = \rho_1 = \rho_2 = 1$ and $a_0, b_0 > 0$ in Theorem 1, we get

Corollary 3. *Let* $P(z) := \sum_{j=0}^{n} \alpha_j z^j$ *be a polynomial of degree n such that* $\alpha_j = a_j + ib_j$ *where* a_j *and* b_j, $j = 0,1,\ldots,n$ *are real numbers. If for some real number* $k \ge 1,$

$$k^n a_n \ge k^{n-1} a_{n-1} \ge \ldots \ge k a_1 \ge a_0 > 0,$$

$$k^n b_n \ge k^{n-1} b_{n-1} \ge \ldots \ge k b_1 \ge b_0 > 0,$$

then all the zeros of $P(z)$ *lie in*

$$|z + k - 1| \le \frac{1}{|a_n|} \left\{ k(a_n + b_n) + 2(k+1) \sum_{j=1}^{n-1} a_j \right\} \tag{3}$$

On combining Theorem B and by taking all coefficients of $P(z)$ to be real in the Corollary 2, the following is immediate:

Corollary 4 *Let* $P(z) := \sum_{j=0}^{n} \alpha_j z^j$ *be a polynomial of degree n, if for some* $k \ge 1,$

$$k^n a_n \ge k^{n-1} a_{n-1} \ge \ldots \ge k a_1 \ge a_0 > 0,$$

then all the zeros of $P(z)$ lie in the intersection of the discs represented by (1) and (3).

PROOF OF THE THEOREM 1

Consider the polynomial

$$
\begin{aligned}
H(z) &= (1-z)P(z) \\
&= -a_n z^{n+1} + (a_n - a_{n-1})z^n + \dots \\
&\quad + (a_{\lambda+1} - a_\lambda)z^{\lambda+1} + (a_\lambda - a_{\lambda-1})z^\lambda + \dots \\
&\quad + (a_1 - a_0)z + a_0 \\
&\quad + i\begin{Bmatrix} -b_n z^{n+1} + (b_n - b_{n-1})z^n + \dots \\ +(b_{\mu+1} - b_\mu)z^{\mu+1} + (b_\mu - b_{\mu-1})z^\mu + \dots \\ +(b_1 - b_0)z + b_0 \end{Bmatrix}
\end{aligned}
$$

$$
\begin{aligned}
&= -a_n z^n(z+k-1) + (ka_n - a_{n-1})z^n \\
&\quad + (ka_{n-1} - a_{n-2} - (k-1)a_{n-1})z^{n-1} \\
&\quad + (ka_{n-2} - a_{n-3} - (k-1)a_{n-2})z^{n-2} + \dots \\
&\quad + (ka_{\lambda+1} - a_\lambda - (k-1)a_{\lambda+1})z^{\lambda+1} \\
&\quad + (ka_\lambda - a_{\lambda-1} - (k-1)a_{n-1})z^\lambda + \dots \\
&\quad + [(a_1 - \rho_1 a_0) + (\rho_1 a_0 - a_0)]z + a_0 \\
&\quad + i\begin{Bmatrix} -b_n z^n(z+k-1) + (kb_n - b_{n-1})z^n \\ +(kb_{n-1} - b_{n-2} - (k-1)b_{n-1})z^{n-1} \\ +(kb_{n-2} - b_{n-3} - (k-1)b_{n-2})z^{n-2} + \dots \\ +(kb_{\mu+1} - b_\mu - (k-1)b_{\mu+1})z^{\mu+1} \\ +(kb_\mu - b_{\mu-1} - (k-1)b_\mu)z^\mu + \dots \\ +[(b_1 - \rho_2 b_0) + (\rho_2 b_0 - b_0)]z + b_0 \end{Bmatrix}
\end{aligned}
$$

$$= z^n \left[\begin{array}{l} a_n(z+k-1)+(ka_n-a_{n-1})+(ka_{n-1}-a_{n-2})\frac{1}{z} \\ +(ka_{n-2}-a_{n-3})\frac{1}{z^2}+\ldots+(ka_{\lambda+1}-a_\lambda)\frac{1}{z^{n-\lambda-1}} \\ +(ka_\lambda-a_{\lambda-1})\frac{1}{z^{n-\lambda}}+\ldots+(a_1-\rho_1 a_0)\frac{1}{z^{n-1}} \\ +(\rho_1-1)a_0\frac{1}{z^{n-1}}+a_0\frac{1}{z^n}-(k-1)\left\{ \begin{array}{l} \frac{a_{n-1}}{z}+\frac{a_{n-2}}{z^2}+\ldots \\ +\frac{a_{\lambda+1}}{z^{n-\lambda-1}}+\frac{a_\lambda}{z^{n-\lambda}} \end{array} \right\} \\ +i\left\{ \begin{array}{l} b_n(z+k+1)+(kb_n-b_{n-1})+(kb_{n-1}-b_{n-2})\frac{1}{z} \\ +(kb_{n-2}-b_{n-3})\frac{1}{z^2}+\ldots+(ka_{\mu+1}-b_\mu)\frac{1}{z^{n-\mu-1}} \\ +(kb_\mu-b_{\mu-1})\frac{1}{z^{n-\mu}}+\ldots+(b_1-\rho_2 b_0)\frac{1}{z^{n-1}} \\ +(\rho_2-1)b_0\frac{1}{z^{n-1}}+b_0\frac{1}{z^n} \end{array} \right\} \\ -(k-1)\left\{ \frac{b_{-1}}{z}+\frac{b_{n-2}}{z^2}+\ldots+\frac{b_{\mu+1}}{z^{n-\mu-1}}+\frac{b_\mu}{z^{n-\mu}} \right\} \end{array} \right]$$

Let $|z|>1$, so that $\frac{1}{|z|^{n-j}}<1, 0\le j\le n$ and we have

$$\begin{aligned} |H(z)| \ge |z|^n \Big[& |\alpha_n||z+k-1| \\ & -\Bigg\{ [ka_n-a_{n-1})+\frac{|ka_{n-1}-a_{n-2}|}{|z|} \\ & +\frac{|ka_{n-2}-a_{n-3}|}{|z|^2}+\ldots+\frac{|ka_{\lambda+1}-a_\lambda|}{|z|^{n-\lambda-1}} \\ & +\frac{|ka_\lambda-a_{\lambda-1}|}{|z|^{n-\lambda}}+\ldots+\frac{|a_1-\rho_1 a_0|}{|z|^{n-1}}+\frac{|\rho_1-1||a_0|}{|z|^{n-1}}+\frac{|a_0|}{|z|^n} \\ & +|k-1|\left(\frac{|a_{n-1}|}{|z|}+\frac{|a_{n-2}|}{|z|^2}+\ldots+\frac{|a_{\lambda+1}|}{|z|^{n-\lambda-1}}+\frac{|a_\lambda|}{|z|^{n-\lambda}} \right) \end{aligned}$$

$$+|kb_n - b_{-1}| + \frac{|kb_{n-1} - b_{n-2}|}{|z|}$$

$$+\frac{|kb_{n-2} - b_{n-3}|}{|z|^2} + \ldots + \frac{|kb_{\mu+1} - b_{\mu}|}{|z|^{n-\mu-1}}$$

$$+\frac{|kb_{\mu} - b_{\mu-1}|}{|z|^{n-\mu}} + \ldots + \frac{|b_1 - \rho_2 b_0|}{|z|^{n-1}} + \frac{|\rho_2 - 1||b_0|}{|z|^{n-1}} + \frac{|b_0|}{|z|^n}$$

$$+|k-1|\left(\frac{|b_{n-1}|}{|z|} + \frac{|b_{n-2}|}{|z|^2} + \ldots + \frac{|b_{\mu+1}|}{|z|^{n-\mu-1}} + \frac{|b_{\mu}|}{|z|^{n-\mu}}\right)\Bigg\}\Bigg]$$

$$> |z|^n \Big[|\alpha_n||z+k-1| - \{[ka_n - a_{n-1})$$

$$+|ka_{n-1} - a_{n-2}| + |ka_{n-2} - a_{n-3}| + \ldots$$

$$+|ka_{\lambda+1} - a_{\lambda}| + |ka_{\lambda} - a_{\lambda-1}| + \ldots$$

$$+|a_1 - \rho_1 a_0| + |\rho_1 - 1||a_0| + |a_0|$$

$$+|k-1|\left(|a_{n-1}| + |a_{n-2}| + \ldots + |a_{\lambda+1}| + |a_{\lambda}|\right)$$

$$+|kb_n - b_{-1}| + |kb_{n-1} - b_{n-2}| + |kb_{n-2} - b_{n-3}|$$

$$+\ldots + |kb_{\mu+1} - b_{\mu}| + |kb_{\mu} - b_{\mu-1}| + \ldots$$

$$+|b_1 - \rho_2 b_0| + |\rho_2 - 1||b_0| + |b_0|$$

$$+|k-1|\left(|b_{n-1}| + |b_{n-2}| + \ldots + |b_{\mu+1}| + |b_{\mu}|\right)\}\Big]$$

$$= |z|^n \Big[|\alpha_n||z+k-1| - \{a_n - \rho_1 a_0 + (1-\rho_1)|a_0|$$

$$+|a_0| + (k-1)\left(\sum\nolimits_{j=\mu}^{n}\left(a_j + |a_j|\right) - |a_n|\right)$$

$$+b_n - \rho_2 b_0 + (1-\rho_2)|b_0| + |b_0|$$

$$+(k-1)\left(\sum\nolimits_{j=\mu}^{n}\left(b_j + |b_j|\right) - |b_n|\right)\}\Big]$$

$$= |z|^n \Big[|\alpha_n||z+k-1| - \{(a_n + b_n) - (\rho_1 a_0 + \rho_2 b_0)$$

$$+(1-\rho_1)|a_0| + (1|-\rho_2)|b_0| + (|a_0| + |b_0|)$$

$$+(k-1)\left(\sum\nolimits_{j=\mu}^{n}\left(a_j + |a_j|\right)\right.$$

$$\left.+\sum\nolimits_{j=\mu}^{n}\left(b_j + |b_j|\right) - |a_n| - |b_n|\right)\}\Big]$$

$$> 0,$$

If

$$|z+k-1| > \frac{1}{|a_n|}\{(a_n+b_n)-(\rho_1 a_0+\rho_2 b_0)$$
$$+(1-\rho_1)|a_0|+(1-\rho_2)|b_0|$$
$$+(|a_0|+|b_0|)+(k-1)\left(\sum_{j=\mu}^{n}(a_j+|a_j|)\right.$$
$$\left.+\sum_{j=\mu}^{n}(b_j+|b_j|)- a_n|-|b_n|\right\}$$

This shows that if $|z|>1$, then $|H(z)|>0$ if

$$|z+k-1| > \frac{1}{|a_n|}\{(a_n+b_n)-(\rho_1 a_0+\rho_2 b_0)$$
$$+(1-\rho_1)|a_0|+(1-\rho_2)|b_0|$$
$$+(|a_0|+|b_0|)+(k-1)\left(\sum_{j=\mu}^{n}(a_j+|a_j|)\right.$$
$$\left.+\sum_{j=\mu}^{n}(b_j+|b_j|)-|a_n|-|b_n|\right\}$$

Hence all the zeros of $H(z)$ with $|z|>1$ lie

$$|z+k-1| \le \frac{1}{|a_n|}\{(a_n+b_n)-(\rho_1 a_0+\rho_2 b_0)$$
$$+(1-\rho_1)|a_0|+(1-\rho_2)|b_0|$$
$$+(|a_0|+|b_0|)+(k-1)\left(\sum_{j=\mu}^{n}(a_j+|a_j|)\right. \tag{1}$$
$$\left.+\sum_{j=\mu}^{n}(b_j+|b_j|)-|a_n|-|b_n|\right\}$$

But those zeros of $H(z)$ whose modulus is less than or equal to 1 already satisfy the inequality (1). Since all the zeros of $P(z)$ are also the zeros of $H(z)$, therefore it follows that all the zeros of $P(z)$ lie in the circle defined by (1) and this completes the proof of Theorem 1.

REFERENCES

1. A. Aziz and B. A. Zargar, Some extensions of Enestrom-Kakeya Theorem, Glasnik Matematički, 31, (1996), 239-244.
2. M. Brambilla, A. Giovannini and R. Ugoccioni, Maps of zeros of the grand canonical partition function in a statistical model of high energy collisions, J. Phys. G: Nucl. Part. Phys., 32((2006), 859-870.
3. R. Fricke, Lehrbuch der Algebra, Friedrich Vieweg and Sohn Verlag, braunschweig, Germany, 1924.
4. N. K. Govil and Q. I. Rahman, On the Eneström-Kakeya Theorem II, Tohoku Math. J. 20, (1968), 126-136.
5. Gulshan Singh and W.M Shah, On the Eneström-Kakeya theorem, Bull. Math. Soc. Sci. Math. Roumanie, Tome 56(104) No. 4, 2013, 505-512.
6. Gulshan Singh and W. M. Shah, On the Eneström-Kakeya theorem, Applied Mathematics, 3(2010), 555-560.
7. W. Heitzinger, W. I. Troch and G. Valentin, Praxis nichtlinearer Gleichungen, Carl Hanser Verlag, munchen, Wien, Germany, Austria, 1985.
8. A. Joyal, G. Labelle and Q. I. Rahman, On the location of zeros of polynomials, Canad. Math. Bull. 10, (1967), 53-63.
9. J. R. Karam, An application of the Kakeya-Eneström theorem, Proc. Int. Conf. Image Sci. in Syst. Technol. CISST, 04(2004), 354-358.
10. M. Marden, Geometry of polynomials, IInd Ed. Math. Surveys 3, Amer. Math. Soc., Providence, R.I. (1966).
11. T. Ooba, On companion systems with state saturation nonlinearity, IEEE Trans. Autom. Control, 50(2003), 1580-1584.
12. Q. I. Rahman and G. Schmeisser, Analytic theory of polynomials, Oxford University Press, Oxford (2002).
13. W. M. Shah and A. Liman, On the zeros of a certain class of polynomials and related analytic functions,Mathematicka Balkanicka, New Series, 19, (2005): Facs., 3-4.
14. W. M. Shah, A. Liman and Shamim Ahmad Bhat, On the Eneström-Kakeya Theorem, International Journal of Mathematical Science, 7, (1-2) (2008), 111-120.
15. T. Sheil-Small, Complex polynomials, Cambridge University Press, Cambridge (2002).
16. J. Shen and G. Strang, Asymptotic analysis of Daubechies polynomials, Proc. Amer. Math. Soc., 124(1996), 3819-3833.

INDEX